Insect Immunity

SERIES ENTOMOLOGICA

VOLUME 48

Insect Immunity

Edited by

J. P. N. Pathak

KLUWER ACADEMIC PUBLISHERS
DORDRECHT / BOSTON / LONDON

ISBN 0-7923-2086-7

Joint edition published by Kluwer Academic Publishers,
P.O. Box 17, 3300 AA Dordrecht, The Netherlands
and
Oxford & IBH Publishing Co. Pvt. Ltd.,
66 Janpath, New Delhi 110001, India.

Kluwer Academic Publishers incorporates
the publishing programmes of
D. Reidel, Martinus Nijhoff, Dr W. Junk and MTP Press.

Sold and distributed in the U.S.A. and Canada
by Kluwer Academic Publishers,
101 Philip Drive, Norwell, MA 02061, U.S.A.

Sold and distributed in India,
Pakistan, Sri Lanka, Bangladesh,
Nepal, Sikkim and Bhutan
by Oxford & IBH Publishing Co. Pvt. Ltd.,
66 Janpath, New Delhi 110001, India.

In all other countries, sold and distributed
by Kluwer Academic Publishers Group,
P.O. Box 322, 3300 AH Dordrecht, The Netherlands.

Printed in India

Preface

Immunity in insects is different from immunity in vertebrates. Insects lack immunoglobulins even though they are capable of reacting against foreign components with effective defense mechanism. There has been a marked advancement in most of the fields of science in the past two decades. Insect immunity is also one of them. It is a developing subject which is now established as a new branch in insect study. This treatise is an attempt to compile meaningful articles of leading workers in this field, nevertheless we do not claim that leadership in insect immunity is by any means restricted to them. The idea is to provide a vibrant description of various aspects of "Insect Immunity". With the rapid development of the subject, it is difficult for any one author to discuss all the aspects of an area in a limited number of pages, even then they have done their utmost to include the entire development of the subject in their articles.

The treatise deals with insect haemocytes, their population, isolation and role in defense mechanism, humoral encapsulation, inducible humoral antibacterial immunity, cellular immune reactions, role of endocrines, role of prophenol oxidase system in cellular communication, haemagglutinins and impact of parasite on insect immune system. Some topics could not be covered because experts in those area though willing could not complete their commitment within time limits.

Variation in approach, style and form is bound to be there in different chapters as no restriction was placed on the authors to organise their views in the manner they liked best. Author was also free to use the terminology, definitions and concepts which he regarded the best. We have no hesitation in admitting that readers may find a different opinions regarding a few topics in the existing literature as I, as the editor, have not interfered with the opinions and expression of the authors.

My gratitude to the authors for their kind cooperation and immense patience. I am grateful to Dr. A.B. Saxena, Professor, School of studies in Zoology, Vikram University, Ujjain and to my colleagues and students whose advice and suggestions were gratefully received.

Last but not least I wish to put on record my gratitude to the University

Grants Commission, New Delhi, as this work was completed during my assignment as Scientist-B with U.G.C.

Scientist B,   J.P.N. PATHAK
Vikram University,
Ujjain, India

Contents

The Contributors

Anna Aspán
Department of Physiological Botany, University of Uppasala, Box 540,
S-75121 Uppasala, Sweden

J. Bahadur
School of Studies in Zoology, Jiwaji University Gwalior, India

Michel Brehélin
Universite de Montpellier II, Place Eugene Bataillon 34095 Montpellier,
France

Y. Carton
Laboratoire de Biologie et Genetique Evolutives, C.N.R.S. 91198, Gif Sur
Yvette, France

Latifa Drif
Universite de Montpellier II, Place Eugene Bataillon 34095 Montpellier,
France

Godwin P. Kaaya
The International Centre of Insect Physiology and Ecology (ICIPE)
P.O. Box 30772, Nairobi, Kenya

A. Nappi
Department of Biology, Loyola University of Chicago, 6525 North Sheridan
Road, Chicago, Illinois 60626, USA

J.P.N. Pathak
Scientist 'B', Department of Zoology, Madhav Science College Vikram
University, Ujjain, India

N.A. Ratcliffe
Biomedical and Physiological Research Group, School of Biological Sciences,
University College of Swansea Singleton Park, Swansea SA2 8PP,
Wales, U.K.

Kenneth Söderhäll
Department of Physiological Botany, University of Uppasala, Box 540,
S-75121 Uppasala, Sweden

M. Sugumaran
Department of Biology, University of Massachusetts, Boston, MA 02125,
USA

Alain Vey
Director Station de Researches de Pathologie Comparee 3038,
Saint-Christol-Lez-Ales, France

S. Bradleigh Vinson
Department of Entomology, Texas A&M University College Station,
TX 77843-2475, USA

CHAPTER 1

Structure, Classification and Functions of Insect Haemocytes

Latifa Drif and Michel Brehélin

Introduction

Since the pioneer work of Swammerdam (1737), who described the haemocytes of *Pediculus humanus*, the blood cells of insects have been the subject of numerous investigations. Until recently, haemocyte classifications were mainly based on the morphology and staining capacities of these cells as observed under light microscopy (Wigglesworth, 1959). Several papers were published and discrepancies in haemocyte classification are obvious. For example, the same haemocyte type may be labelled differently in various insect species or even in the same species, depending on the author of the investigations. Similarly, the same name has been used to identify two (or more) different types of blood cells. This situation gave rise to much confusion among insect haematologists regarding the identification of haemocyte types and considerable disagreement as to the number of blood cell types present in various insect species.

The reasons for such controversy, aside from the diversity of insects (about one million species), are that haemocytes exhibit variable morphologies depending mainly on developmental stages, but also on environmental conditions (diet, temperature etc.). Moreover, some haemocytes are very fragile cells and susceptible to damage or destruction, depending on the methods used for their study.

A study of the ultrastructural features of these cells has provided more details for their determination and new and simpler classifications have been published (Ratcliffe *et al.*, 1985). However, the lack of comparative investigations, taking in account both ultrastructural and functional aspects, has precluded total elimination of the confusion and ambiguities in the terminology used by various authors in their classifications.

The reasons for such a situation were reviewed by Brehelin and Zachary (1986) and a new classification was proposed, which attempted to resolve the controversies. The haemocyte types identification was done in accordance with ultrastructural features and, whenever possible, supported by the functions of these cells in coagulation, endocytosis, encapsulation and wound-healing.

For a better understanding of this classification and of the criteria used for the definition of each type, we propose as a model the orthopteran *Locusta migratoria* (one of the two main species of African migratory locusts). This insect species has been the subject of numerous physiological investigations and the classification of its haemocytes is simple and well established.

Haemocytes of *Locusta migratoria*: GH1, GH2 and GH3

Three of the seven haemocyte types retained in the present classification are found in *L. migratoria*. All are granular haemocytes and found in most of the insect species studied to date. These granular haemocytes are defined and described according to their ultrastructural features and the functions performed by each.

GRANULAR HAEMOCYTES (GH1) (Fig. 1)

GH1 are polymorphous cells with numerous digitations and pinocytic vesicles. The rough endoplasmic reticulum (RER) is always well developed in enlarged cisternae and filled with a flocculent substance. The Golgi complexes are numerous. The main characteristic of GH1 is the presence of granular inclusions which can be divided into three types:

— Inclusions of the first type exhibit an internal structure comprising numerous short microtubules. They originate from the Golgi apparatus.
— Inclusions of the second type are homogeneous electron-dense granules, membrane-bound and also of endogenous origin. They could come from inclusions of the first type by condensation of their content.
— Inclusions of the third type are heterogeneous bodies of a resorptive nature. These membrane-bound granules are filled with vesicles, membranes and amorphous material; they are secondary lysosomes comprising endogenous (primary lysosomes) and exogenous (phagosomes) material. As the size of the microtubular granules varies according the age of the larva, heterogeneous bodies of different size were found in the same cell (Essawy *et al.*, 1985; Delmas, 1988).

Most of the fragile haemocytes which quickly lyse at the time of bleeding are GH1. The precipitation of plasma (Fig. 2) or the formation of veils (Fig. 3) is very common around these lysed cells in *in vitro* studies. The precipitation of plasma helps in the ready identification of GH1 cells in light or phase-contrast microscopy. However, Brehelin and Hoffmann (1980) found that only a part of these cells (about 70%) exhibit such transformation *in vitro*; hence they separated the GH1 population into two groups, i.e., fragile GH1 and stable GH1, which vary from all other granular haemocytes, viz., GH2 and GH3. They are involved in endocytosis and induction of haemolymph coagulation. Their role in coagulation of haemolymph was demonstrated in *Galleria mellonella* (Rowley and Ratcliffe, 1976) and *Locusta migratoria* (Brehelin, 1979). However, in *Locusta migratoria* the process of haemolymph coagulation is also aided by the exocytosis of GH2 (Brehelin, 1979). It should be noted that fragile GH1 have also been reported in *Rhodnius prolixus* where the blood does not coagulate. GH1 are found in numerous insect species, especially lepidopterans, in which they represent the highest percentage among all the haemocytes (Rowley and Ratcliffe, 1981; Essawy *et al.*, 1985); contrarily, they are absent in dipterans. In obsolete literature, GH1 were designated 'coagulocytes' (Gregoire and Florkin, 1950; Hoffmann and Stoeckel, 1968), 'adipohaemocytes' (Beaulaton, 1968), 'plasmalocytes' (Beaulaton, 1968), 'plasmalocytes' (Lai-Fook, 1970; Wigglesworth, 1973) or 'granular haemocytes' (Rowley and Ratcliffe, 1981; Wago, 1982).

GRANULAR HAEMOCYTES 2 (GH2) (Fig. 6)

GH2 are spherical or ovoid cells with an homogeneous cytoplasm and usually a regular surface where digitations or vesicles of pinocytosis are rarely observed. Numerous membrane-bound homogeneous electron-dense granules occur in the cytoplasm. These granules generate by the fusion of the vesicles of the Golgi complex. These cells possess short cisternae (RER) which are never enlarged. In phase-contrast microscopy, these haemocytes are readily recognized by their numerous refringent inclusions (Fig. 5). *In vitro* studies have shown that nearly 50% of the unfixed GH2 cells show exocytosis.

They show little or no phagocytic activity either *in vitro* or *in vivo*. In species in which they are present (*Locusta migratoria*), they help in the process of encapsulation of foreign bodies too large to be phagocytosed. They also help in the process of wound-healing. In the process of encapsulation or wound-healing, they extremely modify as flat cells with desmosome-like junctions. Numerous microtubules also develop in the cytoplasm of these cells during encapsulation. These cells are mainly reported from the larvae of orthopterans (*Locusta migratoria*) and coleopterans (*M. melolontha*) by Brehelin *et al.* (1975). They are absent in most lepidopteran and dipteran species (Zachary and Hoffmann, 1973; Brehelin *et al.*, 1978; Essawy *et al.*, 1985; Brehelin and Zachary, 1986).

The GH2 cells are comparable with the oenocytoids of Wigglesworth (1955), the 'typical granulocytes' of Jones (1962), the 'granulocytes' of Hoffmann and Stoeckel (1968) and 'plasmatocytes' of Rowley and Ratcliffe (1981).

GRANULAR HAEMOCYTES 3 (GH3) (Fig. 7)

These are polymorphic cells with numerous digitations and pinocytic vesicles. The RER is well developed in large cisternae. The Golgi complexes synthesize numerous primary lysosomes in these cells. The cytoplasm includes sometimes a few, but more often numerous heterogeneous bodies. They are considered secondary lysosomes and are rich in acid phosphatase activity. These heterogeneous bodies are considered 'granules' at the light microscopic level; hence these haemocytes were designated granular haemocytes by earlier workers. In light microscopy, GH3 appear as large cells that spread on the glass surface with numerous cytoplasmic expansions (Fig. 7B).

These cells are very active phagocytes cells and engulf all small objects (living and non-living) within reach in the haemocoel of the insect (Fig. 8). Zachary and Hoffmann (1973) noted the numerous occurrence of these cells in the haemolymph of *Calliphora erythrocephala* (Diptera), which actively cleared out tissue debris in the haemolymph at the time of metamorphosis. Brehelin and Boemare (1988) demonstrated that they are the most active cells in recognizing and fixing the bacteria *in vitro*. In addition to dipteran insects, they are also reported from insects of the orders Orthoptera, Coleoptera and Dictyoptera; they are absent, however, in insects of the order Lepidoptera (Brehelin and Zachary, 1986).

Non-granular Haemocytes

For haemocytes of the other types, which are largely non-granular, we maintained in our classification the names given in the literature since, by and large, the authors are in accord concerning the description and designation of each type.

Fig. 1: Granular haemocyte 1 of *Locusta migratoria*. Inclusions of the three kinds are present (arrows) (see text). The granule with an internal structure (one arrow) is close to a Golgi complex (G). Note the enlarged cisternae of the rough endoplasmic reticulum (RER) and the presence of pinocytotic vesicles (arrowhead). Transmission electron microscopy (= TEM). bar = 1 μm

Fig. 2: Transformation of GH1 in *Periplaneta americana* (Dictyoptera) *in vitro*, 5 min after blood removal. The GH1 is surrounded by flocculent material (FM). The GH2 remains intact. TEM. bar = 1 μm

Figs. 3 and 4: Transformations of GH1 of *L. migratoria in vitro*, observed in phase-contrast microscopy 5 min after blood removal. Haemolymph was collected in phosphate buffer saline. Phase-contrast microscopy (= PCM). bar = 7 μm

Fig. 5: GH2 spread on slide in phosphate buffer saline 20 min after blood removal. Note the numerous dense small granules. PCM. bar = 7 μm

Fig. 6: GH2 exhibiting numerous dense homogeneous granules of the same size in their cytoplasm. TEM. bar = 1 μm

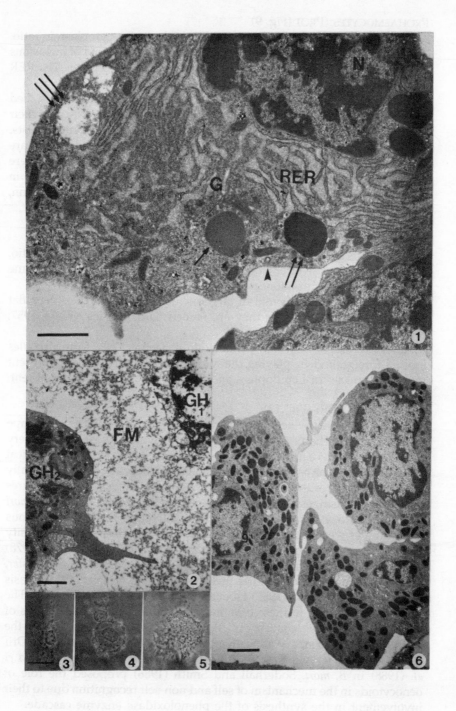

PROHAEMOCYTES (PRO) (Fig. 9)

These are small spherical cells with a high nucleo-cytoplasmic ratio. The thin cytoplasmic layer surrounding the nucleus is devoid of inclusions and mainly characterized by a large amount of free ribosomes. The RER and Golgi apparatus are poorly developed.

Prohaemocytes have been observed in most of the insect species studied to date, but they are rare in the circulatory blood. In light microscopy, their cytoplasm appears basophilic and poorly reactive to histochemical tests.

Because they differ very little and have a high mitotic index, they are considered stem cells for the other haemocytes. In *L. migratoria*, Pro were observed in the haemolymph after perturbations of the haemogram, in which Pro seemed to be released from haemopoietic organs (Hoffmann, 1969).

PLASMATOCYTES (PL) (Fig. 10)

These cells are larger than Pro with a smaller nucleo-cytoplasmic ratio. Their shape is most often regular with very few expansions. In some species, these cells are elongated.

Their cytoplasm is devoid of large inclusions. Due to their very flat shape, in *Drosophila* species they are named 'lamellocytes' (Rizki, 1957; Brehelin, 1982) (Fig. 11).

They exhibit a very low endocytotic capacity. They form capsules around foreign bodies, playing the role of GH2. In fact, they are present mainly in species of Lepidoptera and Diptera, in which GH2 are absent.

OENOCYTOIDS (OE) (Fig. 12)

Oe are large cells which usually exhibit a regular rounded or oval shape with a wide and homogeneous cytoplasm in which ribosomes are very numerous. Their eccentric nucleus contains a large nucleolus. The typical cell organelles, such as mitochondria, Golgi complexes or RER, are usually very rare in these cells. Oenocytoides sometimes exhibit peculiar structures in their cytoplasm, especially in lepidopteran insects, i.e. in *Bombyx mori* (Akai and Sato, 1973) and *Galleria mellonella* (Neuwirth, 1973) numerous microtubules were found while irregular areas with less electron density were reported in the cytoplasm of *Cirphis unipuncta* (Brehelin *et al.*, 1978) and *Heliothis armigera* (Essawy *et al.*, 1985). In *Drosophila melanogaster*, oenocytoids represent crystal-like inclusions in the cytoplasm. On the basis of these crystals, Rizki (1957) labelled them 'crystal cells' (Fig. 13). The presence of tyrosine in these cells indicates their role in the synthesis of phenoloxidase. The concept of their role in the synthesis of the phenoloxidase enzyme was further strengthened by the studies of Drif (1983) in *Aedes aegypti*, Essawy *et al.* (1985) in *H. armigera* and Ashida *et al.* (1988) in *B. mori*. Soderhall and Smith (1986) proposed the role of oenocytoids in the mechanism of self and non-self recognition due to their involvement in the synthesis of the phenoloxidase enzyme cascade.

Oenocytoids are readily recognized under a light microscope due to their large and regular shape, homogeneous cytoplasm and eccentric nucleus. They are found in most of the insects but are absent in Orthoptera (*Locusta migratoria*). They never represent more than 1 to 2% of all the haemocytes present in a normal insect. Another peculiarity of oenocytoids is their labile nature; they are particularly fragile *in vitro* and lyse quickly, ejecting material into the haemolymph.

SPHERULE CELLS (Sph) (Figs. 14 & 15)
These are often very large haemocytes. Their cytoplasm is filled with large-sized inclusions known as 'spherules'. These characteristic inclusions are membrane-bound. In Lepidoptera, they possess an internal structure made of membranes arranged in concentric layers (Akai and Sato, 1973 in *B. mori*; Beaulaton and Monpeyssin, 1977 in *Antheraea pernyi*). In other species (for instance in the coleopteran *Melolontha melolontha*; Devauchelle, 1971) they are electron-dense. In Lepidoptera, their RER is disposed in short but enlarged cisternae and the Golgi complexes are sometimes numerous. Spherules originate from the Golgi apparatus. *In vitro* spherules can be released from the cytoplasm (Fig. 21).

In phase-contrast microscopy (light microscopy), Sph appear irregular in shape, exhibiting an aspect of a 'morula', which facilitates their recognition. They should not be confused either with circulating adipose cells (Drif, 1983; also see Figs. 18-22) or with GH1 or GH3 overloaded with engulfed material (see Fig. 23).

The functions of these cells are unknown. They could play a role in the synthesis of mucopolysaccharides in *B. mori* (Akai and Sato, 1973). Nittono (1960) noted their possible role in silk production in the same species, an hypothesis proposed again by Essawy *et al.* (1985) in another lepidopteran (*H. armigera*). Delmas (1988) noted the absence of Sph in another lepidopteran (*Pieris brassicae*), however, and suggested that this finding supports the hypothesis of Nittono, because in contrast to *B. mori* or *H. armigera*, *P. brassicae* does not spin a cocoon of silk at the time of metamorphosis.

Other Haemocyte Types
Haemocytes that do not belong to the types described above have been observed in some insect species. These comprise the thrombocytoids of Diptera, the haemocytes with numerous small granules in Lepidoptera and Coleoptera and the peculiar granular haemocytes of Hemiptera.

THROMBOCYTOIDS
These are voluminous cells with numerous small mitochondria and dense minute lysosomal granules. The RER is characterized by long parallel and smoothly curved cisternae. But their most prominent feature is more

or less dense and intricate invaginations of the plasmatic membrane intersecting the peripheral cytoplasm. This sometimes leads to the formation of numerous cytoplasmic fragments resembling the platelets of mammals and to the formation of 'naked nuclei', that is, nuclei surrounded by a narrow layer of cytoplasm. Thrombocytoids and their derivatives build up capsules around any large foreign bodies present in the haemocoel (Zachary *et al.*, 1975).

Described for the first time in *Calliphora erythrocephala* by Zachary and Hoffmann (1973), thrombocytoids were observed only in dipteran insects. In our opinion, the 'detached cytoplasmic processes' described by Kaaya and Ratcliffe (1982) in various species of *Glossina* and the 'granular cell fragments' of *Tipula paludosa* (Carter and Green, 1987) belong to the thrombocytoid lineage.

Haemocytes with Numerous Small Granules

The cytoplasm of these cells contains numerous membrane-bound small granules (less than 0.3 μm in diameter) originating from the Golgi complexes. The shape of these cells is often unique, with long (more than 50 μm) cytoplasmic processes. They have been variously labelled as 'podocytes' or 'vermiform cells' in the literature (see Brehelin *et al.*, 1978) and referred to as a special form of plasmatocytes by some authors (Devauchelle, 1971; Essawy *et al.*, 1985). However, they are also present in species in which true plasmatocytes are observed (some Lepidoptera and Coleoptera) and, unlike Pl, do not react in capsule formation. These cells are mainly observed in species of *Prodenia* (Yeager, 1945; Jones, 1959; Pathak, 1989).

Granulocytes of Hemiptera

In *Rhodnius prolixus* and *Triatoma infestans*, the hyaloplasm of some haemocytes, otherwise resembling GH2, is homogeneous and electron-

---

Fig. 7: GH3 of *L. migratoria*. Note the presence of inclusions of different size (7a) and of numerous cytoplasmic expansions (7b).
7a: TEM. bar = 1 μm; 7b: PCM 10 min after blood removal. bar = 10 μm
Fig. 8: GH3 in the process of engulfment of injected latex beads (L). Note the numerous pinocytotic vesicles (arrows) and the Golgi complex (G). N = nucleus. TEM. bar = 1 μm
Fig. 9: Prohaemocyte in *Triatoma infestans* (Hemiptera). This haemocyte shows a high nucleo-cytoplasmic ratio.
9a: TEM. bar = 1 μm; 9b: PCM. bar = 10 μm
Fig. 10: Plasmatocyte in *Galleria mellonella* (Lepidoptera). Note the narrow cisternae of the RER (arrows) and the lack of inclusions. M = mitochondriae.
10a: TEM. bar = 1 μm; 10b: PCM. bar = 10 μm
Fig. 11: Smear of haemolymph from *Drosophila yakuba* after fixation in acid osmic vapours and Giemsa staining. The flat and transparent oval cells are 'lamellocytes' (= plasmatocytes) (arrows). The other cells are GH3 (arrowheads). Light microscopy. bar = 5 μm

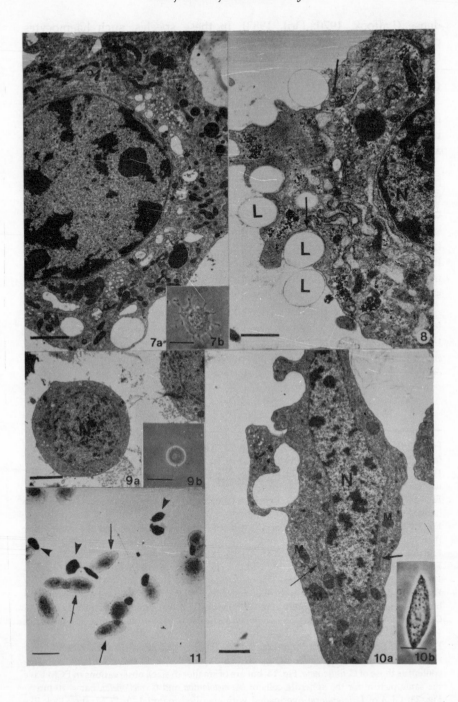

dense (Laifook, 1970; Drif, 1983). In these species, such haemocytes participate neither in encapsulation nor in wound healing, suggesting that they belong to another haemocyte type.

Conclusions

Haemocytes are the main cells involved in the defence mechanism of insects. So their study is of primary importance in understanding the reaction of an insect against a pathogen or the modus operandi whereby a parasite circumvents the defence reactions of its host.

In studying the haemocytes of a new species, we have found that the best approach is: first describe the different types detected under transmission electron microscopy. Next, identify each haemocyte type under light and phase-contrast microscopy. Histochemical tests are useful but inadequate for proper differentiation of the various kinds of granular inclusions occurring in the cytoplasm (Costin, 1975). Let us also emphasize here the deficiency of studies conducted under light microscopy alone; such investigations are unable to distinguish one GH from another and sometimes even from spherule cells.

Functional investigations proved to be relevant in the distinction and identification of insect haemocytes. But classification cannot be determined the basis of functional criteria alone. The best example of this is the case of granular haemocytes, which constitute the fulcrum of most of the controversies.

Fig. 12: Oenocytoid of *Melolontha melolontha* (Coleoptera). Only a few mitochondria (arrowheads) are visible in the voluminous cytoplasm. TEM. bar = 1 μm

Fig. 13: 'Crystal cell' (= Oenocytoid) of *Drosophila melanogaster* (Diptera). Note the large crystal-like inclusions (C). TEM. bar = 1 μm

Fig. 14: Spherule cell of *Cirphis unipuncta* (Lepidoptera). The spherules (S) are large heterogeneous inclusions. TEM. bar = 1 μm

Fig. 15: Spherule cell of *M. melolontha*. The spherules are electron-dense and of a smaller size than in *C. unipuncta*. TEM. bar = 1 μm

Fig. 16: 'Crystal cell' of *D. melanogaster*. Giemsa stain. bar = 15 μm

Fig. 17: 'Crystal cell' of *D. yakuba*. In PCM this cell looks more like a fat body cell (see Fig. 18) rather than the 'Crystal cell' of *D. melanogaster*. bar = 15 μm

Figs. 18 and 19: Fat body cells in the blood of *Aedes aegypti*. These cells are not haemocytes but cells released from the fat body at blood removal. This was proved after a study in TEM. Observations with light microscopy were not sufficient to lead to this conclusion (see Drif, 1983). PCM. bar = 15 μm

Figs. 20-22: Spherule cells of *M. melolontha* (20, 21) and of *G. mellonella* (22). Whereas they exhibit inclusions of very different size in TEM (spherules of *G. mellonella* have the same content as those of *C. unipuncta*, Fig. 14, but are of smaller in size), observations in PCM gave the same picture for the spherule cells of *M. melolontha* and *G. mellonella*. bar = 10 μm

Fig. 23: GH3 of *L. migratoria* overloaded with engulfed material. In PCM, they look like spherule cells (see Figs. 20-22) or fat body cells (see Fig. 19). bar = 15 μm

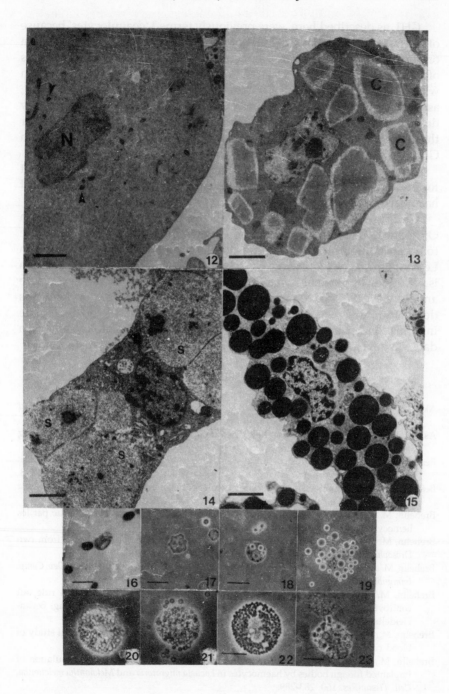

GH1, as described here, were previously termed 'coagulocytes' because of their rapid and strong transformations during blood removal. In orthopteran species, they are characterized by the presence of three types of granules and enlarged cisternae of the RER. But in lepidopterans, in which GH1 exhibit the same ultrastructural features, they do not 'explode', as in orthopterans, but merely slightly degranulate. In Hemiptera, they do not transform at all. Thus the name 'coagulocyte' is inappropriate for these blood cells. Moreover, in some insect species, haemocytes other than GH1 are sometimes able to transform during blood removal.

Similarly, the designation 'granular haemocyte' per se is insufficient for the above-described haemocytes due to the presence of other types of haemocytes with granules.

GH2 are stable *in vitro* and contain electron-dense granules, in Orthoptera, Dictyoptera, Cheleutoptera and Coleoptera. They contribute to encapsulation of foreign bodies too large to be phagocytosed. But in Lepidoptera we never observe GH2. In these species, the capsule formation is performed by plasmatocytes; for this reason, several authors (see Ratcliffe and Rowley, 1981) have labelled our GH2 'granular plasmatocytes'. This designation suggests an analogy between GH2 and Pl, or a possible interrelationship such as the transformation of one type into the other. To date, however, no one has adduced evidence of such a transformation.

REFERENCES

Akai, H. and S. Sato. 1973. Ultrastructure of larval haemocytes of the silkworm, *Bombyx mori* L. (Lepidoptera, Bombycidae). *Int. J. Morphol. Embryol.* 2: 207-231.

Ashida, M., M. Ochiai and T. Niki. 1988. Immunolocalization of prophenoloxidase among haemocytes of the silkworm, *Bombyx mori*. *Tissue and Cell* 20: 599-610.

Beaulaton, J. 1968. Etude ultrastructurale et cytochimique des glandes prothoraciques des vers a sole aux quatrieme et cinquieme ages larvaires. *J. Ultrastruct. Res.* 23: 474-498.

Beaulaton, J. and M. Monpeyssin. 1977. Ultrastructure et cytochimie des hemocytes d'*Antheraea pernyi* Guer. II Cellules, a Spherules et Oenocytoides. *Rev. Biol. Cell.* 28: 13-18.

Brehelin, M. 1979. Mise en evidence de l'induction de la coagulation plasmatique par les hemocytes chez *Locusta migratoria*. *Experientia* 35: 270.

Brehelin, M. 1982. Comparative study of structure and functions of blood cells from two *Drosophila* species. *Cell. Tissue. Res.* 221: 607-615.

Brehelin, M. and D. Zachary. 1983. About insect plasmatocytes and granular cells. *Dev. Comp. Immunol.* 7: 683-686.

Brehelin, M. and D. Zachary. 1986. Insect hemocytes: a new classification to rule out controversy. In: *Immunity in Invertebrates*, (M. Brehelin, ed.). Springer Verlag, Berlin-Heidelberg.

Brehelin, M., D. Zachary and J.A. Hoffmann. 1978. A comparative ultrastructural study of blood cells from nine insect orders. *Cell. Tissue. Res.* 195: 45-47.

Brehelin, M., D. Zachary, J.A. Hoffmann, G. Matz and A. Porte. 1975. Encapsulation of implanted foreign bodies by haemocytes in *Locusta migratoria* and *Melolontha melolontha*. *Cell. Tissue Res.* 160: 283-289.

Brehelin, M. and J.A. Hoffmann. 1980. Phagocytosis of inert particles in *Locusta migratoria* and *Galleria mellonella*: study of ultrastructure and clearance. *J. Insect Physiol.* 26: 103-111.

Brehelin, M. and N. Boemare. 1988. Immune recognition in insects: conflicting effects of autologous plasma and serum. *J. Comp. Physiol. B* 157: 759-764.

Carter, J.B. and E.I. Green. 1987. Haemocytes and granular cell fragments of *Tipula paludosa* larva. *J. Morphol.* 191: 289-294.

Costin, N. 1975. Histochemical observations of the haemocytes of *Locusta migratoria. Histochem. J.* 7: 21-43.

Delmas, J.C. 1988. Adaptation parasitaire de Paecilomyces fumosoroseus (Wize) a l'insecte *Pieris brassicae* (Lep.) et consequences hematologiques de l'infection. These de Doctorat d'Etat, Paris.

Devauchelle, G. 1971. Etude ultrastructurale des hemocytes du Coleoptere *Melolontha melolontha* (L.). *J. Ultrastruct. Res.* 34: 491-516.

Drif, L. 1983. Etude des cellules circulantes d'insectes vecteurs hematophages: ultrastructure et fonctions. These Troisieme Cycle, Montpellier.

Essawy, M., A. Maleville. and M. Brehelin. 1985. The haemocytes of *Heliothis armigera*: Ultrastructure, functions and evolution in the course of larval development. *J. Morphol.* 186: 255-264.

Gregoire, C. and M. Florkin. 1950. Blood coagulation in Arthropods. I. The coagulation of insect blood, as studied with the phase contrast microscope. *Physiologia Comp. Oecol.* 2: 126-139.

Harpaz, I., N. Kislev and A. Zelcer. 1969. Electron microscopic studies on haemocytes of the aegyptian cottonworm *Spodoptera littoralis* (Boisduval) infected with a nuclear polyhedrosis virus, as compared to noninfected haemocytes. I. Noninfected haemocytes. *J. Invertebr. Pathol.* 14: 175-185.

Hayat, M.A. 1973. *Electron Microscopy of Enzymes. Principles and Methods,* 2 vols. Van Nostrand Reinhold, New York and London.

Hoffmann, J.A. 1969. Etude de la recuperation hemocytaire apres hemorragies experimentales chez *Locusta migratoria. J. Insect Physiol.* 15: 1375-1584.

Hoffmann, J.A. and M.E. Stoeckel. 1968. Sur les modifications ultrastructurales des coagulocytes au cours de la coagulation de l'hemolymphe chez *Locusta migratoria. C. R. Acad. Sci.,* D 266: 503-505.

Huxham, I.M. and A.M. Lackie. 1986. A simple visual method for assessing the activation and inhibition of phenoloxidase production by insect haemocytes *in vitro. J. Immunol. Meth.* 94: 271-277.

Jones, J.C. 1959. A phase contrast study of blood cells in *Prodenia* larvae (order Lepidoptera). *Quart. J. Mic. Sci.* 100 (10): 17-23.

Jones, J.C. 1962. Current concepts concerning insect haemocytes. *Amer. Zool.* 2: 209-246.

Kaaya, G.P. and N.A. Ratcliffe. 1982. Comparative study of haemocytes and associated cells of some medically important dipterans. *J. Morphol.* 173: 361-365.

Lai-Fook, J. 1970. Haemocytes in the repair of wound in an insect, *Rhodnius prolixus. J. Morphol.* 130: 297-314.

Neuwirth, M. 1973. The structure of the haemocytes of *Galleria mellonella. J. Morphol.* 139: 105-123.

Nittono, Y. 1960. Studies on the blood cells in the silkworm *Bombyx mori. Bull. Sericult. Exptl. Stat. Tokyo* 16: 171-266. (in Japanese)

Pathak, J.P.N. 1989. Cellular defense mechanism in insects. *Proceedings of the National Symposium on Recent Trends in Immunobiology and Biochemistry,* pp. 31-38. Vikram University, Ujjain India.

Ratcliffe, N.A., A.F. Rowley, S.W. Fitzgerald and C.P. Rhodes. 1985. Invertebrate immunity: basic concepts and recent advances. *Internat. Rev. Cytol.* 97: 183-350.

Rizki, T.M. 1957. The nature of crystal cells of *Drosophila melanogaster. Anat. Rec.* 128: 608.

Rowley, A.F. and N.A. Ratcliffe. 1976. The granular cells of *Galleria mellonella* during clotting and phagocytic reactions *in vitro. Tiss. and Cell.* 8: 437-446.

Rowley, A.F. and N.A. Ratcliffe. 1981. Insects. In: *Invertebrate Blood Cells*, vol. 2 (N.A. Ratcliffe and A.F. Rowley, eds.). Academic Press, London-New York.

Söderhäll, K. and V.J. Smith. 1986. The prophenoloxidase activating system. The biochemistry of its activation and role in arthropod cellular immunity, with special reference to crustaceans. In: *Immunity in Invertebrates* (M. Brehelin ed.). Springer Verlag, Berlin-Heidelberg.

Swammerdam, J. 1737. *Bybel der Nature of Historie der Insekten.* Boerhaave ed.

Wago, H. 1982. Cellular recognition of foreign materials by *Bombyx mori* phagocytes. I. Immunocompetent cells. *Dev. Comp. Immunol.* 6: 591-599.

Wigglesworth, V.B. 1955. The role of the haemocytes in the growth and moulting of an insect, *Rhodunius prolixus* (Hemiptera). *J. Exp. Biol.* 32: 649-653.

Wigglesworth, V.B. 1959. Insect Blood Cells. *Ann. Rev. Entomol.* 4: 1-16.

Wigglesworth, V.B. 1973. Haemocytes and basement membrane formation in *Rhodnius. J. Insect Physiol.* 19: 831-844.

Yeager, J.F. 1945. The blood picture of the sourthern armyworm (*Prodenia eridania*). *J. Agric. Res.* 71: 9-40.

Zachary, D., M. Brehelin and J.A. Hoffmann. 1973. The haemocytes of *Calliphora erythrocephala. Z. Zellforsch.* 141: 55-73.

Zachary, D., M. Brehelin and J.A. Hoffmann. 1975. Role of the "thrombocytoids" in capsule formation in the Dipteran *Calliphora erythrocephala. Cell Tiss. Res.* 162: 343-348.

CHAPTER 2

Haemocytes and Their Population

J. Bahadur

Introduction

The circulatory system in insects is almost entirely of the open type wherein the haemolymph is confined to the body cavity or haemocoel. The dorsal vessel is a closed structure and considered to be the 'primary circulatory pump' (Jones, 1964) or the 'dorsal pulsatile organ' (Jones, 1977). The dorsal vessel, various accessory pulsatile organs and septa help to direct the flow of the haemolymph as well as the body movements of the insect. The haemolymph is the extracellular circulating fluid that fills the body cavity of the insect. It is physically isolated from direct contact with the body tissues by a thin permeable membrane which lines the haemocoel. The haemolymph contains organic and inorganic components in addition to cellular elements termed haemocytes.

Insect haemocytes have been investigated for the last 150 years. Leydig (1859) described them as 'round and spindle-shaped cells' while Magretti (1881) and Cuenot (1895) described them as 'amoeboid cells'. Cuenot classified insect haemocytes into four categories, followed by Hollande (1909, 1911). Wigglesworth (1939, 1959) summarised the earlier classifications and presented one that was widely accepted. Later pioneer workers in the field were Yeager (1945) and Jones (1962). More recently, Gupta (1979) summarized seven main morphologically distinct types of haemocytes: prohaemocytes, plasmatocytes, granulocytes, spherulocytes, oenocytoids, coagulocytes and adipohaemocytes. All types of haemocytes are not readily observed in any one species, at all developmental stages and under all physiological conditions (Gupta, 1979). The structure and classification of haemocytes are described in Chapter 1 of the present volume.

Origin and Evolution of Haemocytes

Dohrn (1876) was the first to describe embryonic blood cells in *Bombyx mori*. Subsequent workers assigned different sites of origin but most authors believe that embryonic haemocytes are mesodermal in origin and are derived from the median part of the inner layer (Mori, 1979). Mori characterized embryonic haemocytes in *Gerris* as prohaemocytes, plasmatocytes and granulocytes and suggested that the plasmatocytes and granulocytes distinctly differ from the prohaemocytes. Many authors have considered mitosis to be responsible for haemocyte multiplication during post-embryonic development but Shapiro (1979a) believed that mitosis alone could not account for the increase in number of haemocytes in the circulatory system. Mitosis has not only been reported in prohaemocytes, but also in other types of haemocytes (Shapiro, 1968; Arnold and Hinks, 1976; Arnold and Sohi, 1974).

In many insects, specialized haemopoietic organs or tissues produce haemocytes during post-embryonic development. Cuenot (1896) was the first to observe such organs as 'phagocytic organs'. There is no agreement among authors about the types of haemocytes that originate from the haemopoietic organs in various insects (Gupta, 1985). Gupta (1985) proposed that granulocytes originate from the prohaemocytes or stem cells through plasmatocytes. Granulocytes perhaps further differentiate into spherulocytes, oenocytoids, coagulocytes and adipohaemocytes. But post-granulocyte differentiation is generally attained through distinct prohaemocytes and plasmatocytes.

Techniques Used in the Study of Haemocytes

(a) *Staining:* Before actual staining, a thin film of the blood is prepared. An excellent method is to dip the insect into water held at 60°C for 1 to 10 minutes, then cut the antenna or leg and place a drop of haemolymph directly over a glass slide, and subsequently draw a cover-slip over it to make a thin film. The blood film is air dried before staining (Arnold and Hinks, 1979).

Alternatively, a drop of formalin (5%) or glutaraldehyde (0.4 M) is placed on a slide, an antennal tip immersed in it and then so cut that the blood comes directly into contact with the solution. The two are stirred well and the mixture then spread on a slide and air dried (Arnold and Hinks, 1979).

Arnold and Hinks (1979) have recommended the following staining methods:

(i) *Giemsa staining:* Immerse the glass slide with the air-dried film in Giemsa solution for 20 minutes to 2 hours (1 drop of concentrate per millilitre distilled water). Rinse the slide in distilled water and then immerse briefly in water to which a few drops of lithium carbonate have been added (this differentiates red-staining structures). Rinse in distilled water again and immerse briefly in distilled water to which a few drops of dilute HCl have been added (this differentiates blue-staining structures). Again rinse the slide in distilled water. Repeat differentiation if staining is too intense. Blot the slide dry and mount in Canada balsam.

(ii) *Haemotoxylin-eosin-alcian blue staining:* Immerse the slide containing an air-dried blood film in 15% acetic acid in methanol for 20 minutes. Then dehydrate the film through grave alcohols and rinse in distilled water. Immerse in 1% alcian blue in 0.1 NHCl for 20 min. Rinse in distilled water and immerse in Harris's haemotoxylin for 20 min. Again rinse the slide in distilled water, differentiate the blood film in acid alcohol and subsequently blue in Scott's solution. Pass the slide through an ascending series up to 90% alcohol and stain in 1% alcohol eosin (prepared in 90% alcohol). Rinse off excess stain, dehydrate the blood film in absolute alcohol, then clear in xylene and lastly mount the slide in Canada balsam.

(b) *Total haemocyte count (THC) determination:* THC was first determined by Tauber and Yeager (1934, 1935) in several insects. The physiological saline used contained: NaCl—4.65 g, KCL—0.15 g, $CaCl_2$—0.11 g, gentian violet—0.005 g and acetic acid—0.125 ml/100 ml. Over the years this solution has been slightly modified by various workers (Shapiro, 1979b). But the method has remained fairly consistent over the past 25 years, namely, drawing haemolymph into a Thoma white blood cells pipette, diluting it in the ratio of 1:100 in the physiological saline solution (Gupta and Sutherland, 1968), counting the THC in a haemocytometer and determining the THC per 1 mm^3 of blood according to the formula suggested by Jones (1962).

(c) *Differential haemocyte count (DHC) determination:* DHC is the relative number of different types of haemocytes and is determined per fixed number of haemocytes counted (say 200 per slide). The method of Shapiro (1966) is considered typical and consists of counting in fixed smear preparations. Counting can also be done in histological sections (Nappi and Stoffolano, 1972a) and hanging-drop preparations.

(d) *Hanging-drop preparation for THC and DHC:* Gupta (1979) gave an identification key for the haemocyte type under a phase-contrast microscope in hanging-drop preparations of fixed and unfixed haemolymph. The procedure is as follows:

i) On a square cover-slip place a tiny drop of saline-versene (NaCl—0.9 g, KCl—0.942 g, $CaCl_2$—0.082 g, $NaHCO_3$—0.002 g, distilled water—100 ml + 2% versene).

ii) Cut the tip of the antenna or leg (or proleg) and let a drop of haemolymph flow into the saline-versene drop.

iii) Carefully turn the cover-slip upside down and place it over the depression or cavity slide. Seal the sides of the cover-slip with petroleum jelly.

iv) Examine the hanging-drop preparation under the phase-contrast microscope.

v) The haemocytes are more evenly distributed near the periphery and can be focussed.

(e) *Absolute haemocyte count (AHC) determination:* AHC represents an estimate of both the THC and DHC in relation to the blood volume (BV) and is estimated by multiplying the THC by the BV (Gupta, 1985).

(f) *Blood (haemolymph) volume determination:* A dye solution method to determine the blood volume (BV) was employed by Yeager and Munson (1950), which was later modified by Lee (1961). Shapiro (1979b) calculated the blood volume in *Galleria* by the following formula:

$$V = \frac{d(c' - c'')1}{c''}$$

where V = BV in microlitres; d = volume of amaranth dye (1% aqueous solution) injected in microlitres; c' = original concentration of dye (percentage); c'' = concentration of the dye after circulation (percentage). To obtain the blood volume, divide the V value by the body weight of the insect.

The haemogram may show great variability within the same insect and hence caution must be exercised in determination and interpretations. The variability may be due to differences in techniques, imprecision in use, physiological state of the insect, its age and sex (internal factors) and external factors, such as disease, starvation, wounds etc.

Haemocyte Population Changes During Development

Changes in the total haemocyte count (THC) during growth and development of insects have been reported by a number of workers. But it may vary with treatment of the insects and developmental and physiological conditions (Yeager, 1945; Wheeler, 1963; Jones, 1967a; Shapiro, 1979a). Salt (1970) further cautioned that the THC may vary due to sampling procedures and inherent changes in the ratio of circulating to sedentary haemocytes. The changes that take place in the population of haemocytes during development in several orders of insects are briefly reviewed on following pages.

(a) *Orthoptera:* THC was studied in a number of orders by Tauber and Yeager (1935) who reported that total haemocyte counts were higher in females of *Periplaneta* and *Blatta* carrying ootheca than in normal females. They also demonstrated that adults of cricket (*Udeopsylla*), grasshopper (*Melanoplus*), dragonfly (*Plathermis*) and stink bug (*Euschistus*) had a higher haemocyte population than their nymphs and thus interpreted that a gradual increase in haemocyte number takes place during development. In *Schistocerca* (Mathur and Soni, 1937) and in *Locusta* (Webley, 1951) also the adults have a higher cell population than the nymphs. In *Periplaneta* the THC is reduced at the time of ecdysis but a very rapid build-up of haemocytes takes place within 24 hr of ecdysis (Patton and Flint, 1959). Smith (1938), however, reported variations in cell counts in different individuals of *Periplaneta* and thereby showed a bimodal frequency as compared to trimodal frequency distribution of cells counts in *Gryllus*.

(b) *Hemiptera:* Salt (1970) summarized the data of Tauber and Yeager (1935) on total haemocyte counts in 62 species of insects and concluded that, in general, immature forms and adults show noticeable variation in haemocyte counts. In stage V nymphs of *Oncopeltus* (Feir and O'Connor, 1969) the THC increased slightly during the first three days after ecdysis, then decreased during the next three days followed by increase. Feir (1964) had earlier reported an increase in THC after ecdysis from nymph IV to V and adult of *Oncopeltus*, followed by reduction in number 48 hr post-ecdysis. But Feir and O'Connor (1969) considered these differences due to variation in haemolymph volume and/or differences in the adherence of haemocytes to tissues or to wounds. Wigglesworth (1955) found that THC increased prior to moult in *Rhodnius*, followed by decrease at ecdysis and again increase after moult. But in *Halys dentata* (Hemiptera), Bahadur and Pathak (1971) demonstrated an abrupt decrease in THC after ecdysis, a rise during mid-stadium and a slight fall prior to the next ecdysis. They showed an increase in THC throughout the life cycle with the adult's being maximum.

In the stage V nymph of *Rhodinus*, the haemocytes were quite high compared to the stage IV nymph (Jones, 1967b). Jones and Liu (1961) reported that prior to moult, the prohaemocytes decreased in number while the adipohaemocytes increased. At moult, the plasmatocytes and oenocytoids decreased and after moult, the plasmatocytes and oenocytoids increased but the granulocytes decreased. Jones (1967b) found great differences in the haemocyte population between fed and unfed nymphs on the fifth day after ecdysis. In fed ones, the plasmatocytes and oenocytoids increased greatly while in adult females of *Rhodnius*, following a blood meal, the prohaemocytes increased but lysing granulocytes decreased. Saxena and Tikku (1990) studied the effect of plumbagin a phytochemical, on the haemocytes of *Dyrdercus koenigii* after topical application. Scanning electron microscopical studies showed deformity in surface morphology in almost all the five types of haemocytes categorized in the bug, especially that of granular haemocytes and plasmatocytes, which are devoid of their fillopods in the treated insects. The fat droplets of adipohaemocytes shrink while oenocytoids are affected to a lesser degree. They also noted the variation in total and differential haemocyte counts performed after 24 and 48 hours of treatment. The population of all the haemocytes was reduced in the treated insects.

(c) *Coleoptera:* Tauber and Yeager (1936) reported higher THC in the larvae of *Phyllophya* and *Osmorderma* than their adults. But in *Leptinotara* (Arvy *et al.*, 1948) no differences in THCs were observed among larvae, pupae and adults. Jones (1950) found plasmatocytes and granulocytes to be the principal haemocytes in *Tenebrio* larvae throughout the larval period, Clark and Chadbourne (1960) observed no differences in THCs of diapausing and non-diapausing larvae and pupae of the pink bollworm. But Raina and Bell (1974) noted significant reduction in all cell types during diapause compared to the non-diapause state in the pink bollworm. They found adipohaemocytes to be three times higher in number in pharate pupae that had developed from diapausing larvae than in those from non-diapausing ones.

(d) *Diptera:* Nappi (1970) found only plasmatocytes in the 1st instar larvae of *Drosophila euronotus* but in the 2nd instar both oenocytoids and lamellocytes. In the 3rd instar larvae, the plasmatocytes decreased about 50% while the oenocytoids and lamellocytes increased 25% each. During the prepupae period the plasmatocytes, oenocytoids and lamellocytes occurred in nearly equal proportions. In the first larvae of *Calliphora erythrocephala*, haemocytes appeared at the end of this larval stage but the number continued to increase during the larval period. Similarly, in the larvae of *Orthellia caesrion*, very few haemocytes occurred in circulation in a two-day-old larva but the number increased during the larval period (Nappi and Stoffolano, 1972a).

In *Musca domestica* and *Musca autumnalis* larvae, the haemocytes are located in the posterior part of the body (Nappi and Stoffolano, 1972b) but during the larval period they increase in number. However, at the time of pupation they were found to be distributed throughout the haemocoel.

In *Sarcophaga bullata*, prohaemocytes were found in all the larval stages, though rarely (Jones, 1956, 1967c). Plasmatocytes were present in all stages of larvae and pupae but occurred exclusively in adults. However, the population of plasmatocytes decreased during development whereas that of granulocytes increased. During the first six days of larval life, the spherulocytes increased in proportion but decreased abruptly afterwards. During the last larval stage, THC increased prior to pupation while in the prepupal to brown pupal stage, THC decreased and the granulocytes disappeared from circulation prior to adult emergence.

(e) *Lepidoptera:* In *Prodenia eridania*, THC did not vary significantly in 6th instar larvae as they approached pupation and the cell numbers were higher in larvae than in adults (Rosenberger and Jones, 1960). In the larvae of *Trichoplusia ni* (Laigo and Paschke, 1966), the cell counts varied while in *Heliothis zea* (Shapiro *et al.*, 1969) little change in number was observed in six- to ten-day-old larvae. In *Heliothis virescens* (Vinson, 1971), THC increased during the larval period. Moulting 5th instars of the armyworm *Pseudaletia unipuncta* had higher THCs but after moult the cell population decreased and continued to decline until prepupation (Wittig, 1965). But in the larvae of *Euxoa declorata* the cell number increased during the 2nd to 6th instars (Arnold and Hinks, 1976). Shapiro (1979a) could arrive at no generalized conclusion or trend because of conflicting data for *Prodenia, Trichoplusia, Pseudaletia* and *Euxoa*. No significant differences between THCs of non-diapause larvae and five-day-old pupae of *Pectinophora gossypiella* (pink bollworm) were observed by Clark and Chadbourne (1960). Though the prohaemocyte, adipohaemocyte and coagulocyte populations varied from stage to stage, the authors considered these variations were due to haemocyte functions in food transport, storage and metabolism.

In the Mediterranean flour moth, the haemocyte population increased during larval development and peaked in the prepupal period, but during pupation the number decreased and again increased slightly on the first day of adult life (Arnold, 1952). The THC peaked at every moult in the larva of *Bombyx mori* but the highest THC was attained in the 5th instar (Nittono, 1960). It later declined and reached its lowest level after adult emergence. During the active growth period in each instar, plasmatocytes occurred in high numbers but decreased in the 5th instar, increased again before adult eclosion and attained a maximum value in adults. Though the granulocytes and adipohaemocytes also peaked at each moult, yet minimal values were recorded in adults. After mid-stage of pupation, larval prohaemocytes and spherulocytes were not observed though the latter were detected in adults. The number of oenocytoids usually decreased after pupation.

Lea (1964) recorded large variations in cell number from the 4th instar to the adult in *Hyalophora cecropia* but at pupation the counts decreased and remained low in newly emerged adults. Plasmatocytes predominated in 4th and earlier instars. While plasmatocytes and granulocytes made up more than 90% of the total haemocyte population, oenocytoids were relatively scarce and spherulocytes occurred at the time of cocoon spinning.

Jones (1967a) concluded that THCs increase during larval development and the counts from heat-fixed larvae were significantly higher than those from unfixed larvae. The data suggested that cell counts increase continuously during cocoon spinning but decrease at the prepupal stage.

In general, it has been reported that during development the different types of haemocytes behave differently. In *Prodenia* (Yeager, 1945), *Galleria*

(Shapiro, 1966) and *Euxoa* (Arnold and Hinks, 1976), the prohaemocytes are reduced during the last larval instar up to prepupation when the number again increases.

On the other hand, the granulocytes increase in number during the larval period (Yeager, 1945; Shapiro, 1966; Nittono, 1960; Arnold and Hinks, 1976).

The adipohaemocytes increase during larval development in *Galleria* (Jones, 1964; Shapiro, 1966). However, in *Hyalophora cecropia* diverse types of adipohaemocytes occur. Type I occurs in high number in larvae and type II in prepupae and newly moulted pupae through the first day of adult life (Lea and Gilbert, 1966).

Cyclical Changes

(a) *Due to Ecdysis:* In *Periplaneta americana*, Patton and Flint (1959) reported a significant decrease in THC at moult, which began 48 hours prior to moulting and became more significant during the next 24 hours. Changes in haemocyte population after moult also occurred in *Locusta* (Webley, 1951), *Rhodnius* (Wigglesworth, 1955), *Sarcophaga* (Jones, 1956), *Bombyx mori* (Nittono, 1960) and *Periplaneta* (Wheeler, 1963). In *Rhodnius* (Jones, 1967c), no consistent changes in the proportion of prohaemocytes, plasmatocytes, granulocytes occurred at ecdysis, but after moult the plasmatocytes increased in number while the granulocytes decreased. Tauber (1935, 1937) noted a decrease in the proportion of mitotically dividing cells before and during moult in *Blatta orientalis* and an increase after ecdysis, particularly on the third day.

In *Schistocerca* (Lee, 1961), the decrease in THC at moult was attributed to an increase in blood volume and coagulability of blood. But Jones (1956) and Arnold (1966) attrtibuted the decrease in THC to adherence of haemocytes to tissues. In the bug *Halys dentata*, Bahadur and Pathak (1971) demonstrated an abrupt fall in THC after ecdysis, a rise during mid-stadium and a slight fall before the next ecdysis. They showed that THC increases throughout the life cycle and the adult has the highest cell counts.

(b) *Due to blood volume:* Total haemocytes counts (THCs) are represented by total counts per cubic millilitres of haemolymph and thus any change in volume of water in the haemolymph would affect THC though it might exert no effect on the absolute number of haemocytes. In *Bombyx mori* larvae, a constant blood volume occurred up to the 11th day after spinning, then fell and maintained a constant level from the 12th day up to the time of eclosion, when another reduction took place (Florkin, 1937). Nittono (1960) showed an increase in THC at each moult with no reduction in blood volume and thereby inferred that the increase was due to a real increase in circulating haemocytes as well as due to a reduction in blood volume. An inverse relationship between the THC and the blood volume in *Bombyx mori* larvae was reported by Nittono (1960) and also in the last stage nymphs and adults at ecdysis in *Periplaneta* (Wheeler, 1962; 1963). Jones (1956) concluded that in *Sarcophaga* changes in THC could not be related to changes in blood volume. In *Galleria* larva (Shapiro, 1966), THC increased with no decrease in blood volume and thus it was concluded that THC is/was not a reflection of a decrease in blood volume. In spite of change in THC and blood volume, the absolute number of circulating haemocytes increased from 2.2 million to 4 million.

In *Locusta migratoria*, Webley (1951) attributed the higher concentration of haemocytes in older adults to a decrease in blood volume rather than to an increase in the number of haemocytes. In *Periplaneta* the absolute number of haemocytes remained the same in spite of changes in blood volume and THC (Wheeler, 1962, 1963).

Jones (1967b) found that in *Rhodnius*, the plasmatocytes increased and the granulocytes decreased in number during the fasting period following each ecdysis. In the 4th and 5th instars, the plasmatocytes decreased while the granulocytes increased after a meal while in adults, following a meal, the plasmatocytes increased and the granulocytes decreased in number. Shapiro (1979a) suggested that changes in haemocyte population and types of haemocytes are better appreciated when the THC, DHC and BV measurements are combined.

(c) *Due to periodicity:* Changes in haemocyte populations during the light-dark cycle were studied in the giant cockroach, *Blaberus giganteus*, by Arnold (1969). He reported changes in the proportions of spherulocytes and granulocytes during the light-dark cycle in 25% of the insects but found no change in the proportion of prohaemocytes and plasmatocytes. He could not demonstrate convincing evidence for a diurnal response but noted that 'there is a potential for periodicity in the haemocyte complex and a need to consider it in the planning of experiments'. Changes in THC in the 6th instar larvae of *Pseudaletia* at different times during the third day after moult were associated with feeding rhythm rather than diurnal rhythm (Wittig, 1966). Jones (1967d) found that blood volume increased in *Rhodnius* after feeding and the proportion of granulocytes and oenocytoids also increased while the plasmatocytes decreased. Jones (1967d) concluded that the blood picture is affected by the nutritional state of the insect.

Haemocyte Population Changes due to Implantation of Endocrine Glands

Crossley (1965) indicated that changes in haemocytes at the time of puparium formation in *Calliphora* may be under hormonal control. Jones (1967b) showed the involvement of the anterior endocrine glands of *Rhodnius* in the alternation of the blood picture. He showed that if the 5th instar *Rhodnius* is ligated behind the head before moulting, hormones are fully secreted, the amount of haemolymph becomes abnormally high and the number of prohaemocytes and plasmatocytes is influenced. Ligature after the critical period did not effect such changes. Thus Jones (1967b) concluded that hormones regulate the amount of haemolymph and the number of haemocytes.

Crossley (1968) influenced the titre of moulting hormones by ligature in *Calliphora* and also injected β-ecdysone in feeding larvae and found that the haemocyte population was modified by a significant increase in proportion to circulating phagocytic cells. He ligated the larvae into an anterior part containing the Weismann's ring and a posterior part. The latter could not pupate and maintained a low percentage of phagocytic cells. But when β-ecdysone was injected into the posterior part, the percentage of phagocytic cells increased. His histogram indicated the interconversion in the haemocyte population due to the titre of hormone and thus Crossley claimed that the endocrine glands and hormones are involved in maintaining the blood picture.

In *Locusta migratoria migratorioides*, electrocoagulation of the pars intercerebralis in the male increased the blood volume and reduced the haemocyte population (Hoffmann, 1970). It was found that the change in blood volume might alter the absolute number but not the population of each type of haemocyte. Electrocoagulation of the pars intercerebralis in the female increased THC abruptly with a significant increase in number of plasmatocytes. When the corpora allata were extirpated, a decease in THC occurred but when implanted, the THC increased considerably. Implantation of the corpora cardiaca decreased blood volume, thereby increasing the concentration of haemocytes. Extirpation of the corpora affected the rate of differentiation of haemocytes (Hoffmann, 1970).

Hinks and Arnold (1977) divided the last instar larva of *Euxoa declarata* into three compartments with ligatures, the anteriormost bearing the endocrine glands and two pairs of haemopoietic organs. The enhancement of mitosis in the haemopoietic organs and increase in the number of prohaemocytes and plasmatocytes in the anteriormost compartment suggested the roll of endocrine glands. Though the author could not identify the hormone responsible for these changes, yet they have suggested that the changes were induced by ecdysone.

In *Halys dentata*, Pathak (1983) showed with extirpation experiments that the hormonal secretion of the corpora allata influenced THC up to the age of six days but when the corpora cardiaca were extirpated, the count decreased significantly after six days. Implantation experiments led the author to conclude that the corpora allata influence the count during the early part of adult life while the secretion of the neurosecretory cells of the brain and corpora cardiaca influence THC during the later part of adult life.

Essawy *et al.* (1984) while working with the larvae of *Heliothis armigera* hypothesised that the plasmatocytes contain a substance, perhaps from the brain, which modifies the permeability of the prothoracic gland for exchanges between the gland and the haemocytes. They indicated that plasmatocytes and coagulocytes are influenced by the titre of ecdysone.

Pathak (1986) concluded that the endocrine glands influence the total and differential haemocyte counts in various ways, viz.: (a) interconversion of haemocytes, (b) increasing the mitotic index, (c) releasing the haemocytes from haemopoietic tissues and (d) mobilizing sessile and adhering cells. He emphasized that there is an interrelationship between blood volume and haemocyte population and that the blood volume is controlled by several factors, one of them being hormonal. The diuretic hormone or its analogue 5-hydroxytryptamine may influence blood volume suddenly, resulting in an increase in THC.

It was inferred that in *Halys dentata*, brain and corpora cardiaca extracts of hydrated six-day-old adults were able to decrease the blood volume due to the presence of the diuretic hormone (Pathak, 1989) and hence THC increased. An extract of the thoraco-abdominal ganglionic mass of dehydrated insects possesses predominantly the antidiuretic hormone, which increased the blood volume but reduced the total haemocyte count of the recipients.

In *Halys dentata*, the THC and number of absolute circulating cells were increased significantly by a corpora allata extract of both hydrated and dehydrated insects but no effect on the blood volume was recorded. It was therefore concluded that an increase or decrease in the THC is not merely a reflection of decrease or increase in the blood volume. Probably in *Halys dentata*, the injection of corpora allata extract or juvenile hormone had some impact on the physiological status of the insect which, in turn, stimulated the non-circulating haemocytes into circulation to enhance the THC, and absolute circulating haemocytes or sessile haemocytes were induced by the higher juvenile hormone titre into circulation for some specific function (Pathak, 1991).

Effect of Diseases and Micro-organisms on Haemocyte Population

Arnold (1952) observed no significant changes in THC in *Anagasta* larvae infected by a microsporidium protozoan. But in *Trichoplusia* (Laigo and Paschke, 1966) and in *Apis* (Gilliam and Shimanuki, 1967), THC decreased after the infection of *Nosema* while in *Adelina sericesthis* (Weiser and Beard, 1959) it increased under the influence of different protozoan infections. The haemocytes of *Galleria* multiplied in greater number following inoculation with a trypanosome (Zotta and Teodoresco, 1933). However, the authors concluded that the haemocytic response was not specific to the protozoan.

Tauber and Yeager (1935, 1936) and Beard (1945) noted an increase in THC in the haemolymph of insects due to bacterial infection. But many authors have observed drastic reduction in the number of haemocytes during bacterial infections (Kostritsky *et al.*, 1924; Babers, 1938; Wittig, 1965). In bacteria-infected *Prodenia* larvae, haemocyte counts either remained unchanged or declined markedly; thus Rosenberger and Jones (1960) concluded that the haemocytes are not very effective in protecting the insect. In *Galleria*, THC increased up to 48 hours following infection with a *Staphylococcus* and decreased rapidly (Werner and Jones, 1969). Changes in THC of the haemolymph of *Galleria* as a result of bacterial infection, have been followed by a number of workers including Werner and Jones (1969). They all noted a decrease in plasmatocytes and an increase in granulocytes, prohaemocytes and spherulocytes. Metalnikov (1927) had earlier concluded that each microbe provoked a specific reaction in the haemocytes. In *Anagasta*, Arnold (1952) found a large number of adipohaemocytes associated with bacterial infection. He noted reduction in plasmatocytes and increase in adipohaemocytes after the infection of *Sarcina lutea*. Werner and Jones (1969) found that spherulocytes and oenocytoids did not change in proportion following infection. But high doses of infection in *Pseudaletia* larva resulted in a drastic reduction in plasmatocytes and a great increase in spherulocytes and prohaemocytes with no change in the number of granulocytes (Wittig, 1966). Low doses, on the other hand, had a different effect on the blood picture because the granulocytes and plasmatocytes decreased while the spherulocytes and prohaemocytes increased significantly (Wittig, 1966). Wille (1974) observed no effect of infection on the haemocyte morphology of *Apis*.

Not much information is available on the changes in haemocytes due to fungal infection. Speare (1920) observed no adverse effect on the haemocytes of *Prodenia*. But in some cases the haemocytes formed giant cells (Boezkowska, 1935) or rounded up (Cermakova and Samsinakova, 1960) or showed degenerative changes leading to lysis (Sussman, 1952). Sirotina (1961) found profound pathological changes in the blood of *Leptinotarsa* following infection with *Beauveria*, and reported marked increase in the number of degenerating cells and drastic decrease in the number of phagocytic haemocytes. But in *Cecropia* the blood cells increased in number following infection with the fungi *Aspergillus flavus* and *A. niger*. However, the cell number decreased as the fungus multiplied.

Wittig (1962) reviewed the pathology of insect blood cells during viral infection. Following infection with a virus in *Pseudaletia*, the granulocytes initially increased, then decreased while the spherulocytes and prohaemocytes increased; nonetheless there was an overall drastic reduction in THC (Wittig, 1968). But Xeros and Smith (1954) reported a great increase in THC in *Tipula* during nucleopolyhedrosis infection. In *Galleria* larva, infection with the nucleopolyhedrosis virus caused a reduction in THC only 10 days later, when the plasmatocytes decreased and the granulocytes increased significantly. Adipohaemocytes were lost and degenerating cells and spherulocytes increased (Shapiro, 1967, 1968). In *Heliothis zea* larva, the THC decreased drastically at a high virus concentration; at a low concentration there was no decrease but the spherulocytes increased (Shapiro *et al.*, 1969). Shapiro concluded that 'qualitative or quantitative changes in haemocytes are non-specific responses caused by the presence of foreign substances and/or stress conditions'.

Rickettsiosis in *Melolontha*, caused by *Rickettsia melolontha*, effectively reduced the plasmatocytes to one-sixth their normal value and the adipohaemocytes to one-third. There was no melanization in the haemolymph (Krieg, 1958). Recently, Pathak and Soni (1990) studied the impact of diseases on the unfixed total haemocyte counts of *Bombyx mori* during development. They noted an abrupt decrease in THC in naturally infected (diseased) worms. They also found a noticeable decrease in THC in laboratory infected worms. They pointed out that haemocytes struggle against the micro-organisms in infected worms by means of phagocytosis and nodule formation.

Conclusions

The insect haemolymph contains prohaemocytes, plasmatocytes, granulocytes, spherulocytes, oenocytoids, coagulocytes and adipohaemocytes as morphologically distinct haemocytes but all the types are not observed in any one species. The embryonic haemocytes are mesodermal in origin, derived from the middle part of the inner layer as prohaemocytes, plasmatocytes and granulocytes. Further post-granulocyte differentiation into other types takes place during post-embryonic development through prohaemocytes, plasmatocytes and granulocytes.

The haemocytes are generally stained by Giemsa and haemotoxylin-eosin-alcian blue in fixed smears. The hanging-drop procedure is utilized for unfixed preparations under phase-constrast microscopy. Total haemocyte count (THC) is determined according to the method of Gupta and Sutherland (1968). Differential haemocyte count (DHC) is determined by the method of Shapiro (1966). The blood volume (BV) can likewise be determined by Shapiro's (1979b) method.

During insect development the haemocyte population undergoes change in THC, DHC and BV. These changes have been briefly discussed in Orthoptera, Hemiptera, Coleoptera, Diptera and Lepidoptera. The haemocyte population is also affected by ecdysis, periodicity and blood volume changes.

Various extirpation and transplantation experiments in different insects have demonstrated that the endocrine glands do have a profound affect on haemocyte number and their differentiation.

Haemocyte population and specific types of haemocytes are affected by protozoan and fungal diseases as well as those caused by bacteria and viruses.

REFERENCES

Arnold, J.W. 1952. The haemocytes of the Mediteranean flour moth, *Ephstia kuhniella* Zell. (Lepidoptera: Pyralididae). *Can. J. Zool.* 30: 352-364.

Arnold, J.W. 1966. An interpretation of the haemocyte complex in a stonefly, *Acroneuria arenosa* (Plecoptera: Perlidae). *Can. Entomol.* 98: 394-411.

Arnold, J.W. 1968. An interpretation of the haemocyte complex in a stonefly, *Acroneuria arenosa* (Plecoptera: Perlidae). *Can. Entomol.* 98: 394-411.

Arnold, J.W. 1969. Periodicity in the proportion of haemocyte categories in the giant cockroach, *Blaberus giganteus*. *Can. Entomol.* 101: 68-77.

Arnold, J.W. and C.F. Hinks. 1976. Haemopoiesis in Lepidoptera. I. The multiplication of circulating haemocytes. *Can. J. Zool.* 54: 1003-1012.

Arnold, J.W. and C.F. Hinks. 1979. Insect haemocytes under light microscopy: techniques. In: *Insect Haemocytes* (A.P. Gupta, ed.). Cambridge University Press, Cambridge.

Arnold, J.W. and S.S. Sohi. 1974. Haemocytes of *Malacosoma disstria* Huebner (Lepidoptera: Lasiocampidae): Morphology of the cells in fresh blood and after cultivation *in vitro*. *Can. J. Zool.* 52(4): 481-485.

Arvy, L., M. Gebe and J. Lhoste. 1948. Contribution a l'etude morphologique du sang de *Chrysomela decemlineata* Say. *Bull. Biol. Fr. Belg.* 82: 37-60.

Babers, F.H. 1938. A septicemia of the southern armyworm caused by *Bacillus cereus*. *Ann. Entomol. Soc. Amer.* 31: 371-373.

Bahadur, J. and J.P.N. Pathak. 1971. Changes in the total haemocyte counts of the bug, *Halys dentata*, under certain conditions. *J. Insect Physiol.* 17: 329-334.

Beard, R.L. 1945. Studies on the milky disease of Japanese beetle larvae. *Conn. Agric. Exp. Stn. Bull.* 491: 505-583.

Boezkowska, M. 1935. Contribution a l'etude de le immunite chez les cheniller de *Galleria mellonella* L. contre les champignons entomophytes. *C. R. Soc. Bio.* 119: 39-40.

Cermakova, A. and A. Samsinakova. 1960. Ueber den mechanisums des derchdringens des pilzes *Baauveria bassiana* Vuill. in die larvae von *Leptinotarsa decemliceata* Say. *Ceak. Parasitol.* 7: 231-236.

Clark, E.W. and D.S. Chadbourne. 1960. The haemocytes of nondiapause and diapause larvae and pupae of the pink bollworm. *Ann. Entomol. Soc. Amer.* 53: 682-685.

Clark, R.M. and W.R. Harvey. 1965. Cellular membrane formation by plasmatocytes of diapausing *Cercropia* pupae. *J. Insect Physiol.* 11: 161-175.

Collin, N. 1963. Les haemocytes de la larva de *Melolontha melolontha* L. (Coleoptera: Scarabaeidae). *Rev. Pathol. Veg. Entomol. Agric. Fr.* 42: 161-167.

Crossley, A.C. 1965. Transformation in the abdominal muscles of the blue blowfly *Calliphora erythrocephala* (Meig) during metamorphosis. *J. Embryol. Exp. Morphol.* 14: 89-110.

Crossley, A.C. 1968. The fine structure and mechanism of breakdown of larval intersegmental muscles in the blowfly *Calliphora erythrocephala*. *J. Insect Physiol.* 14: 1389-1407.

Cuenot, L. 1895. Etude physiologiques sur les Orthopteres. *Arch. Biol.* 14: 293-341.

Dohrn, A. 1876. Notizen zur Kenntnis der Insektenetwicklung. *Z. Wiss. Zool.* 26: 122-138.

Essawy, M., A. Mleville and M. Brehelin. 1984. Evolution of haemogram during the larval development (last instar) of *Heliothis armigera*. Invertebr. Immunol. Conf. 17-29. Sept. 1984, Montpellier.

Feir, D. 1964. Haemocyte counts of the large milkweed bug, *Oncopeltus fasciatus*. *Nature* (Lond.) 202: 1136-1137.

Fier, D. and G.M. O'Connor, Jr. 1969. Liquid nitrogen fixation: A new method for haemocyte counts and mitotic indices in tissue sections. *Ann. Entomol. Soc. Amer.* 62: 246-249.

Fisher, R.A. 1935. The effect of acetic acid vapor treatment on the blood cell count in the cockroach *Blatta orientalis* L. *Ann. Entomol. Soc. Amer.* 28: 146-153.

Florkin, M. 1937. Variations de composition du plasma sanguin au cours de la metamorphose du ver a soie. *Arch. Int. Physiol.* 60: 17-31.

Gilliam, M. and H. Shimanuki. 1967. *In vitro* phagocytosis of *Nosema apis* spores by honeybee haemocytes. *J. Insect Physiol.* 12: 1369-1375.

Gupta, A.P. 1979. Identification key for haemocyte types in hanging drop preparations. In: *Insect Haemocytes*, pp. 227-229 (A.P. Gupta, ed.). Cambridge University Press, Cambridge.

Gupta, A.P. 1985. Cellular elements in the haemolymph. In: *Comparative Insect Physiology, Biochemistry and Pharmacology*, pp. 401-451 (G.A. Kerkut and L.I. Gilbert, eds.). Pergamon Press, Oxford-New York.

Gupta, A.P. and D.J. Sutherland. 1968. Effects of sublethal doses of chlordane on the haemocytes and midgut epithelium of *Periplaneta americana*. *Ann. Ent. Soc. Amer.* 61(4): 910-918.

Hinks, C.F. and J.W. Arnold. 1977. Haemopoiesis in Lepidoptera. II. The role of the haemopoietic organs. *Can. J. Zool.* 55: 1740-1755.

Hoffmann, J.A. 1970. Regulations endocrines de la production et de la differenciation des hemocytes chez un insecte. 1. Orthoptere: *Locusta migratoria*. *Gen. Comp. Endocrinol.* 15: 198-219.

Hollande, A.C. 1909. Contribution al'etude du sang des Coléoptères. *Arch. Zool. Exp. Gen.* (Ser. 5) 21: 271-294.

Hollande, A.C. 1911. Etude histologiques comparee du sang des insects des a hemorrhee et des insects sans l'hemorrhee. *Arch. Zool. Exp. Gen.* (Ser. 5) 6: 283-323.

Jones, J.C. 1950. Cytopathology of the hemocytes of *Tenebrio molitor* Linnaeus (Coleoptera). Ph.D. Thesis. Iowa State College, Ames, Iowa.

Jones, J.C. 1956. The hemocytes of *Sarcophaga bullata* Parker. *J. Morphol.* 99: 233-57.

Jones, J.C. 1962. Current concepts concerning insect hemocytes. *Amer. Zool.* 2: 209-246.

Jones, J.C. 1964. Differential hemocyte counts from unfixed last stage *Galleria mellonella*. *Amer. Zool.* 4: 337.

Jones, J.C. 1967a. Changes ;in the haemocyte picture of *Galleria mellonella* (Linnaeus). *Biol. Bull.* (Woods Hole) 132: 211-221.

Jones, J.C. 1967b. Effect of repeated haemolymph withdrawals and of ligaturing the head on differential counts of *Rhodnius prolixus* Stal. *J. Insect. Physiol.* 13: 1351-1360.

Jones, J.C. 1967c. Estimated changes within the haemocyte population during the last larval and early pupal stages of *Sarcophaga bullata* Parker. *J. Insect Physiol.* 13: 645-646.

Jones, J.C. 1967d. Normal differential counts of haemocytes in relation to ecdysis and feeding in *Rhodnius prolixus* Stal. *J. Insect Physiol.* 13: 1133-1141.

Jones, J.C. 1977. The circulatory system of insects. Thomas, Springfield, Illinois.

Jones, J.C. and D.P. Liu. 1961. Total and differential hemocyte counts of *Rhodnius prolixus* Stal. *Bull. Entomol. Soc. Amer.* 7: 166.

Kostritsky, M., M. Toumanoff and S. Metalnikov. 1924. *Bacterium tumefaciens* chez la chenille de *Galleria mellonella*. *C.R. Acad. Sci.* 179: 225-227.

Krieg, A. 1958. Weitere untersuchungen zur Pathologie der Rickettrise von *Melolontha* spec. *Z. Naturforsch.* 13b: 374-379.

Laigo, F.M. and J.D. Paschke. 1966. Variations in total haemocyte counts as induced by a nosemosis in the cabbage looper, *Trichoplusia ni*. *J. Invertebr. Pathol.* 8: 175-179.

Lea, M.S. 1964. A study of the hemocytes of the silkworm *Hyalophora cecropia*. Ph.D. Thesis. Northwestern University, Evanston, Illinois.

Lea, M.S. and L.I. Gilbert. 1961. Cell division in diapausing silkworm pupae. *Amer. Zool.* 1: 568-569.

Lea, M.S. and L.I. Gilbert. 1966. The hemocytes of *Hyalophora cecropia*. *J. Morphol.* 118: 197-216.

Lee, R.M. 1961. The variation of blood volume with age in the desert locust (*Schistocerca gregaria* Forsk.). *J. Insect Physiol.* 6: 36-51.

Leydig, L. 1859. Zur Anatomie der Insekten. *Arch. Anat. Physiol. Med.* 1859, 33-89.

Magretti, P. 1881. Del prodotto di secrezione in alcuni meloidi esame microscopico. *Boll. Sci.* 1: 23-27.

Mathur, C.B. and B.N. Soni. 1937. Studies on *Schistocerca gregaria* Forsk. IX. Some observations on the histology of the desert locust. *Indian J. Agric. Res.* 7: 317-325.

Metalnikov, S. 1927. L'infection microbienne et l'immunite chez la milte de abeiles *Galleria mellonella*. Pasteur Institute, Masson, Paris.

Mori, H. 1979. Embryonic hemocytes: Origin and development in insect hemocytes., pp. 3-27 (A.P. Gupta, ed.). Cambridge University Press, Cambridge.

Nappi, A.J. 1970. Hemocytes of Larvae of *Drosophila euronotus* (Diptera: Drosophillidae). *Ann. Entomol. Soc. Amer.* 63: 1217-1224.

Nappi, A.J. and J.G. Stoffolano, Jr. 1972a. Haemocytic changes associated with the immune reaction of nematode-infected larvae of *Orlshellia caesarion. Parasitology* 65: 295-302.

Nappi, A.J. and J.G. Stoffolano, Jr. 1972b. Distribution of haemocytes in larvae of *Musca domestica* and *Musca autumnalis* and possible chaemotaxis during parasitisation. *J. Insect Physiol.* 18: 169-179.

Nittono, Y. 1960. Studies on the blood cells in the silkworm, *Bombyx mori* L. *Bull. Seric. Esp. Stn.* (Tokyo) 16: 171-266.

Pathak, J.P.N. 1983. Effect of endocrine glands on the unfixed total haemocyte counts of the bug, *Halys dentata. J. Insect Physiol.* 29: 91-94.

Pathak, J.P.N. 1986. Haemogram and its endocrine control in insects. In: *Immunity in Invertebrates*, pp. 49-59 (M. Brehelin, ed.). Springer-Verlag, Berlin-Heidelberg.

Pathak, J.P.N. 1989. Hormonal control of water regulation in *Halys dentata* (Hemiptera). *Acta. Entomol. Bohemoslov.* 86: 33-38.

Pathak, J.P.N. 1991. Effect of endocrine extracts on the blood volume and population of haemocytes in *Halys dentata* (Pentatomidae: Heteroptera). *Entomol.* 16: 251-255.

Pathak, J.P.N. and A.K. Soni. 1990. Impact of diseases on the unfixed total haemocyte counts of *B. mori* L. during development, pp. 13-26. In: *Recent Trends in Sericulture* (J. Promada Kumari, ed.). Tirupati University, Tirupati, India.

Patton, R.L. and R.A. Flint. 1959. The variation in the blood cell counts of *Periplaneta americana* (L). during a molt. *Ann. Ent. Soc. Amer.* 52: 240-242.

Raina, A.K. and R.A. Bell. 1974. Haemocytes of the pink bollworm *Pectinophora gossypiella* during larval development and diapause. *J. Insect Physiol.* 20: 2171-2180.

Rosenberger, C.R. and J.C. Jones. 1960. Studies on the total blood counts of the southern armyworm larva, *Prodenia eridania* (Lepidoptera). *Ann. Entomol. Soc. Amer.* 55: 351-355.

Salt, G. 1970. *Cellular Defense Reactions of Insects.* Cambridge University Press, Cambridge.

Saxena, B.P. and K. Tikku. 1990. Effect of plumbagin on haemocytes of *Dysdercus koenigii* F. *Proc. Indian Acad. Sci.* 99: 119-124.

Shapiro, M. 1966. Pathologic changes in the blood of the greater wax moth, *Galleria mellonella* (Linnaeus) during the course of starvation and nucleopolyhedrosis. Ph.D. Thesis. University of California, Berkeley, California.

Shapiro, M. 1967. Pathologic changes in the blood of the greater wax moth, *Galleria mellonella*, during the course of nucleopolyhedrosis and starvation. I. Total haemocyte count *J. Invertebr. Pathol.* 9: 111-113.

Shapiro, M. 1968. Pathologic changes in the blood of the greater wax moth, *Galleria mellonella*, during nucleopolyhedrosis and starvation. II. Differential haemocyte counts. *J. Invertebr. Pathol.* 10: 230-234.

Shapiro, M. 1979a. Changes in haemocyte population. In: *Insect Haemocytes.*, pp. 475-523 (A.P. Gupta, ed.). Cambridge University Press, Cambridge.

Shapiro, M. 1979b. Techniques for total and differential haemocyte counts and blood volume, and mitotic index determinations. In: *Insect Haemocytes*, pp. 539-561 (A.P. Gupta, ed.). Cambridge University Press, Cambridge.

Shapiro, M., R.D. Stock and C.M. Ugnoffo. 1969. Haemocyte changes in larvae of the bollworm, *Heliothis zea*, infected with a nucleopolyhedrosis virus. *J. Invertebr. Pathol.* 14: 28-30.

Sirotina, M.I. 1961. Haematological detection of microbiological measures taken against the Colorado beetle. *Dolk. Akad. Nauk. USSR* 140: 720-723.

Smith, D.S. 1968. *Insect Cells: Their Structure and Function*. Oliver and Boyd, Edinburgh.

Smith, H.W. 1938. The blood of the cockroach, *Periplaneta americana* L. Cell structure and degeneration and cell counts. Studies of contact insecticides XIII. *Tech. Bull.*, no. 71., New Hampshire Agric. Exp. Stn. Durham.

Speare, A.T. 1920. Further studies of *Sorosporella uvella*, a fungus parasite of noctuid larvae. *J. Agric. Res.* 18: 399-439.

Sussman, A.S. 1952. Studies on an insect mycosis. III. Histopathology of an aspergillosis of *Platysamia cercropia* L. *Ann. Entomol. Soc. Amer.* 45: 233-245.

Tauber, O.E. 1935. Studies on insect hemolymph with special reference to some factors influencing mitotically dividing cells. Ph.D. Thesis. Iowa State College, Ames, Iowa.

Tauber, O.E. 1937. The effect of ecdysis on the number of mitotically dividing cells in the hemolymph of the insect *Blatta orientalis*. *Ann. Entomol. Soc. Amer.* 30: 35-39.

Tauber, O.E. and J.F. Yeager. 1934. On the total blood (haemolymph) cell count of the field cricket *Gryllus assimillis pennsylvanicus* Burm. *Iowa State Coll. J. Sci.* 9: 13-24.

Tauber, O.E. and J.F. Yeager. 1935. On the total blood counts of insects. I. Orthoptera, Odonata, Hemiptera and Homoptera. *Ann. Entomol. Soc. Amer.* 28: 229-240.

Tauber, O.E. and J.F. Yeager. 1936. On the total hemolymph (blood) cell counts of insects. II. Neuroptera, Coleoptera, Lepidoptera and Hymenoptera. *Ann. Entomol. Soc. Amer.* 29: 12-18.

Vinson, S.B. 1971. Defense reaction and haemocytic changes in *Hellothis virecens* in response to its habitual parasitoid *Cardiochiles nigriceps*. *J. Invertebr. Pathol.* 18: 94-100.

Webley, D.P. 1951. Blood cell counts in the African migratory locust (*Locusta migratoria migratorioides* Reiche and Faimaire). *Proc. Roy. Ent. Soc. Lond.* A26: 25-37.

Weiser, J. and R.L. Beard. 1959. *Adelina sericesthis* n. sp., a new coccidian parasite of scarabalis larvae. *J. Insect Pathol.* I: 99-106.

Werner, R.A. and J.C. Jones. 1969. Phagocytic haemocytes in unfixed *Galleria mellonella* larvae. *J. Insect Physiol.* 15: 425-437.

Wheeler, R.E. 1962. Changes in hemolymph volume during the moulting cycle of *Periplaneta americana* L. (Orthoptera). *Fed. Proc.* 21: 123.

Wheeler, R.E. 1963. Studies on the total hemocyte count and hemolymph volume in *Periplaneta americana* (L.) with special reference to the last moulting cycle. *J. Insect Physiol.* 9: 223-235.

Wigglesworth, V.B. 1939. *The Principles of Insect Physiology*, 1st ed. Mehuen. London.

Wigglesworth, V.B. 1955. The role of the haemocytes in the growth and moulting of an insect, *Rhodnius prolixous* (Hemiptera). *J. Exp. Biol.* 32: 649-663.

Wigglesworth, V.B. 1959. Insect blood cells. *Ann. Rev. Ent.* 14: 1-16.

Wille, H. 1974. Studies on the haemolymph of *Apis mellifera* L. 5. Relationships between the morphology of the leukocytes and four disease elements. *Nilt. Schweiz. Entomol. Ges.* 47: 133-149.

Wittig, G. 1962. The pathology of insect blood cells. A review. *Amer. Zool.* 2: 257-273.

Wittig, G. 1965. Phagocytosis by blood cells in healthy and diseased caterpillars. I. Phagocytosis of *Bacillus thuringiensis* Berliner in *Pseudaletia unipuncta* (Haworth). *J. Invertebr. Pathol.* 7: 474-488.

Wittig, G. 1966. Phagocytosis by blood cells in healthy and diseased caterpillars. II. A consideration of method of making haemocyte counts. *J. Invertebr. Pathol.* 8: 461-477.

Wittig, G. 1968. Phagocytosis by blood cells in healthy and diseased caterpillars. III. Some observations concerning virus inclusion bodies. *J. Invertebr. Pathol.* 10: 211-229.

Xeros, N. and K.M. Smith. 1954. Further studies on the development of viruses in the cells of insects. *Proc. Int. Conf. Electron Microsc. (Lond.) Viruses*: 259-262.

Yeager, J.F. 1945. The blood picture of the southern armyworm (*Prodenia eridania*). *J. Agric. Res.* 71: 1-40.

Yeager, J.F. and Munson, S.C. 1950. Blood volume of the cockroach *Periplaneta americana* determined by several methods. *Arthropoda* I: 225-265.

Zotta, G. and A.M. Teodoresco. 1933. Formule leucocytaire de le chenille de *Galleria mellonella* infectes par le *Leptomonas pyrrhocoris*. *C.R. Soc. Biol.* 114: 314-316.

CHAPTER 3

Isolation of Pure Populations of Insect Haemocytes

N. A. Ratcliffe

Introduction

Most living systems consist of mixed cell populations which interact in complex ways to maintain the functional integrity of the whole organism. Such interactions are particularly well illustrated in the vertebrate immune system in which, for example, macrophages and T- and B-lymphocytes co-operate intimately to mediate the cellular and humoral immune defences.

The classical method of studying cell-to-cell co-operation is to isolate and characterize the component cells before utilizing them for functional assays in various combinations. For example, the use of cell separation techniques has led to a much greater understanding of lymphocyte functioning (Gupta and Good, 1980) whilst such techniques are also most important for investigating the differentiation of peripheral blood cells from the bone marrow (Ali, 1986).

Evidence is now accumulating that cell-to-cell co-operation is also a vital component of invertebrate immunity (Brillouet et al., 1984; Ratcliffe et al., 1984; Söderhäll et al., 1986; Johansson and Söderhäll, 1989; Kobayashi et al., 1990; Anggraeni and Ratcliffe, 1991). Thus, the axial organ cells of the starfish, Asterias rubens, can be stimulated to produce 'antibody-like molecules' by the co-operation of non-adherent, adherent and phagocytic cells (Brillouet et al., 1984). In crustaceans, Söderhäll and his colleagues have produced a model for communication between different haemocyte types in the cellular defences. The semi-granular cells firstly degranulate in response to non-self materials and liberate components of the prophenoloxidase system and a 76kDa protein into the plasma. These factors, but particularly the 76kDa protein, then initiate degranulation of the granular cells and the activation of the semi-granular, granular cells and also, possibly, the hyaline cells, to phagocytose or encapsulate foreign invaders of the haemocoel (Söderhäll and Smith, 1986; Johansson and Söderhäll, 1989). Such cell communication is probably basically similar in insects since the initial attachment of degranulating granular cells on non-self surfaces has also been recorded (Ratcliffe and Gagen, 1977) and, more recently, it has been shown that granular cells provide some factor(s) which is necessary for maximal phagocytosis by the plasmatocyte cell type (Anggraeni and Ratcliffe, 1991).

However, in working with crustacean and insect haemocytes, the important point to emphasize is that progress in our understanding of cell-to-cell communication in the immune systems of these animals has been accelerated with the development of techniques for separating the component cells (Söderhäll and Smith, 1983; Mead *et al.*, 1986).

Earlier Attempts to Isolate Insect Haemocytes

A number of techniques for obtaining relatively pure populations of insect blood cells have already been described. These techniques utilized differential adhesion of the haemocytes (Bohn, 1977; Chain and Anderson, 1982), selective depletion following immunization (Chain and Anderson, 1982), discontinuous gradient centrifugation (Peake, 1979; Cook *et al.*, 1985; Huxham and Lackie, 1988; Farkăs, 1991), and continuous gradient centrifugation (Mead *et al.*, 1986; Anggraeni and Ratcliffe, 1991).

DIFFERENTIAL ADHESION

Bohn (1977) utilized the differential adhesive properties of the haemocytes of the cockroach, *Leucophaea maderae*, to produce nearly homogeneous monolayers of either plasmatocytes or granular cells. Haemocyte suspensions were placed in a tissue culture chamber enclosing two cover-slips held 1 mm apart. The chamber was subsequently inverted at various times in order to distinguish adherent and non-adherent cells. The plasmatocytes were highly adhesive and enriched monolayers of these cells formed on the top cover-slip while the granular cells slid downwards and attached mainly to the lower cover-slip.

Chain and Anderson (1982), working with the haemocytes of the wax moth, *Galleria mellonella*, also obtained highly enriched preparations of plasmatocytes by allowing harvested cells to attach and spread on glass cover-slips for only 10-15 minutes before vigorously washing off the non-adherent cells. Only the plasmatocytes attached in such a short time; all the other cell types were washed away. This technique has the advantage of being fairly simple to perform although it is possible that defence responses might be triggered during adhesion by degranulation of many of the labile granular cells/cystocytes. Furthermore, the cells attach in monolayers and are therefore difficult to manipulate and test in different combinations.

SELECTIVE DEPLETION FOLLOWING IMMUNIZATION

Chain and Anderson (1982) combined the different adhesion techniques described above for obtaining enriched plasmatocyte preparations, with an immunization procedure which, 1 hour after the injection of 9×10^5 bacteria, resulted in almost pure supensions of granular cells in the haemocoel. This technique might well be useful even though the granular cells obtained may already have been activated by exposure to the bacteria and may therefore behave atypically *in vitro*. The presence of bacterial components might also affect the subsequent reactions of the cells.

DISCONTINUOUS GRADIENT CENTRIFUGATION

Peake (1979) detailed the methodology for separating and characterizing the haemocytes of the dipteran, *Calliphora vicinia*. Discontinuous gradients of Ficoll (Sigma) were set up and cells up to 1×10^6 analyzed following centrifugation from 0.5 to 4.0 hours at 5,000 g max. With this technique, a high percentage (75-85%) of prohaemocytes and phagocytes were recovered while most of the other cell types, such as oenocytoids (20-30% recovery) and thrombocytoids (barely a few recovered) were almost completely lost by lysis or adhesion to the walls of the tubes. In addition, Peake's protocol includes resuspension of the haemocytes with a Pasteur pipette by 'repeated suction'. Since many dipterans do not contain the fragile cystocyte or coagulocyte cell type which induces plasma gelation, such a resuspension technique is possible. Treatment of haemocytes from most other insect orders in such a manner would induce a massive degranulation of the cells and an immediate gelation of the plasma. Furthermore, these processes would be accelerated with Peake's experimental method as it fails to utilize either anticoagulants or endotoxin-free conditions.

More recently, Cook *et al.* (1985) and Huxham and Lackie (1988) have utilized discontinuous gradients of Percoll (Pharmacia) to purify the spherule cells of the lepidopterans, *Heliothis virescens* and *Malacosoma disstria*, and a small population of granular cells in the locust, *Schistocerca gregaria*, respectively. The spherule cells were purified on two-step gradients of 55% and 84% Percoll and formed a band at the interface which could easily be harvested. The spherule cell band was >95% pure with the other cell types remaining aggregated or coagulated at the top of the gradient. Cook *et al.* (1985) believed that the incorporation of additional Percoll steps would facilitate the isolation of other cell types but the problem of the extensive aggregation and coagulation of these cells was not addressed. Huxham and Lackie (1988) set up discontinuous gradients composed of 25%, 35%, 45% and 65% Percoll which, in contrast to Cook *et al.* (1985), were made up in an anticoagulant solution. After centrifugation, however, at 1,200 g for 15 minutes at 4°C, they only separated one cell population which comprised just 5% of the total haemocyte numbers. These cells were highly granular, 95% pure, and contained most of the phenoloxidase activity of the total blood cell population. Clearly, it is useful to be able to separate such subpopulations of cells but the Cook *et al.* (1985) and Huxham and Lackie (1988) procedures are both suboptimal if large numbers of different pure haemocyte populations are required for studies on cell-to-cell co-operation *in vitro*. Huxham and Lackie (1988) claimed to have attempted, without success, to separate the *Schistocerca* blood cells using continuous Percoll gradients. They attributed this failure to the nature of the locust haemocytes, which comprise a finely graded range of granular cells that frustrate efforts to isolate subpopulations.

A remarkably simple and successful separation of the blood cells of the greater wax moth, *Galleria mellonella*, utilizing a single step discontinuous gradient was recently reported by Farkás (1991). The separation procedure involved layering the haemolymph of 10-12 last-instar larvae onto a single layer of Ficoll-Paque (Pharmacia; although Histopaque from Sigma would also probably suffice) precooled to 2-4 °C. The insects had previously been injected with an anticoagulant, as detailed in Ratcliffe *et al.* (1986), and the haemolymph was collected in 0.5 to 1.0 ml of the same solution. The optimal separation ratio of haemolymph: Ficoll-Paque was 3.0 cm : 2.4 cm in the centrifuge tube. The tubes were then centrifuged in a swing-out rotor at 500 g at 2-4 °C for 10-15 minutes. This resulted in an 87-91% pure plasmatocyte layer at the haemolymph-Ficoll-Paque interface and a 90-96% pure granular cell layer at the bottom of the tube. This technique is discussed in more detail later and compared with the successful separation method of Mead *et al.* (1986) described below.

CONTINUOUS GRADIENT CENTRIFUGATION

Mead *et al.* (1986) and Anggraeni and Ratcliffe (1991) utilized continuous gradients of Percoll and, like Farkăs (1991), were successful in separating the plasmatocytes and granular cells of *Galleria mellonella* (Fig. 1). Since these two cell types are the main ones responsible for immunoreactivity in this insect species and, since they have been isolated in high numbers, this procedure, like that of Farkăs (1991), is an important step in providing pure cell subpopulations for testing *in vitro*. In addition, Mead *et al.* (1986) have shown that the protocol developed is also applicable to other insect species, including the cockroach, *Blaberus craniifer*, and the tobacco hornworm, *Manduca sexta*. The description given below is a detailed consideration of some of the problems in cell separation using continuous gradient techniques based on Percoll and discontinuous gradients as described by Farkăs (1991), and provides protocols not only for purifying, but also for utilizing separated subpopulations in cell-to-cell co-operative studies. Some of the problems involved have been discussed in earlier publications (Mead *et al.*, 1986; Ratcliffe *et al.*, 1986).

Problems in Isolating Insect Haemacytes and Their Use in Subsequent Experiments

Experience has shown that the vast diversity of insects is reflected in the nature of their blood cells, with each species presenting a different array of problems to be overcome for successful separation of the various haemocyte types. For example, Cook *et al.* (1985) derived a successful protocol for purifying the spherule cells of two lepidopteran species, *Heliothis virescens* and *Malacosoma disstria*, but were unable to separate this cell type in a third lepidopteran, *Orgyia leucostigma*, by the same method. Even more remarkable are failed attempts in our laboratory to purify the haemocytes of the cockroach, *Blaberus discoidalis*, using a protocol which successfully separated the blood cells of the closely related species, *Blaberus craniifer* (Mead *et al.*, 1986). Under the light microscope, not only do the haemocytes of these species look identical, but they also cross-react with monoclonal antibodies (Mullet *et al.*, unpublished observations). Possibly, as noted in *S. gregaria* by Huxham and Lackie (1988), the *B. discoidalis* cells may comprise a more finely graded range of granule-containing cells than those of *B. craniifer*, which makes separation on continuous gradients impossible.

Problems involved with isolating pure haemocyte populations of insects and their subsequent utilization include those associated with bleeding the animals, the separation technique, harvesting the cells and obtaining functional preparations of the isolated cells.

BLEEDING

Many insects, but particularly vector species such as mosquitoes, sandflies, etc., are very small and contain only limited amounts of blood for cell separation. It might be possible to overcome this problem, as outlined below, by pooling the blood from a number of insects.

Animals should also be surface-sterilized with 70% alcohol before bleeding. The bleeding technique adopted depends very much on the size and type of insect to be used. Thus, many species, such as *Galleria mellonella*, *Manduca sexta* and *Blaberus craniifer*, can be bled by leg amputation or puncturing (Mead *et al.*, 1986). Price and Ratcliffe (1974) have provided a useful survey of bleeding techniques for a range of insect species.

The almost instantaneous coagulation and gelation of the haemolymph following bleeding is also another difficulty to be overcome in many species (Fig. 2). This coagulation reaction is due to the fragile nature of a granular cell type referred to as the granular cell, cystocyte or coagulocyte in insects. These cells are rapidly activated to discharge their contents by contact with minute quantities of microbial products, such as endotoxins and β 1-3-glucans (Leonard *et al.*, 1985; Ratcliffe *et al.*, 1991), so that all glassware and solutions should, as far as possible, be free of these substances. Treating glassware at 180 °C for 2-4 hours and washing plastic equipment in E-Toxa-

Clean (Sigma) is usually sufficient to remove endotoxin. All the solutions to be used should also be made up in endotoxin-free water (Travenol Labs Ltd.) and filter sterilized. Such extreme precautions are probably essential with most insect species whilst they are apparently unnecessary with crustacean blood cells (Söderhäll and Smith, 1983).

The literature contains an array of techniques and solutions for preventing haemolymph coagulation in insects (reviewed in Ratcliffe *et al.*, 1986), including the use of glacial acetic acid vapour, chilled insects and injected anticoagulants. In our experience, insects should be chilled on ice for 5-10 min to restrict movements and to slow down haemolymph coagulation.

All insect haemocytes, therefore, have to be stabilized with an anticoagulant before placing on gradients for separation. The most efficacious anticoagulant available is based on citric acid and EDTA, which not only reduce free Ca^{++}, which is probably essential for the coagulation process, but also impart a low pH to stabilize the cells (Fig. 2). A typical anticoagulant solution would consist of 0.098 M NaOH, 0.186 M NaCl, 0.017 M EDTA (free acid) and 0.041 M citric acid at pH 4.5 and osmolality of 440 m $OsmKg^{-1}$ for *Galleria* larvae. For other species, such as *M. sexta* and *B. craniifer*, the osmolality of the anticoagulant solution can simply be adjusted with NaCl (Mead *et al.*, 1986). Farkąs (1991) also saturated the anticoagulant with phenylthiourea; however, this is not recommended for haemocytes to be used for subsequent functional studies on the prophenoloxidase system as this complex is inhibited by this chemical. Another choice to be made at this stage is whether to inject the anticoagulant into the insect prior to bleeding or to bleed the insect directly into the anticoagulant solution. Recently, Anggraeni and Ratcliffe (1991), using continuous Percoll gradients, improved the separation of *Galleria* plasmatocytes from the 67% purity obtained by Mead *et al.* (1986) to 94.5% by bleeding directly into the anticoagulant solution in a cold room at 4°C (Table 1). Prior injection may well result in wounding reactions which

Table 1: Percentage purity of plasmatocytes, granular cells and spherule cells from *Galleria mellonella* before and after separation on continuous Percoll gradients[a]

Haemocyte types	Before separation[b]	After separation (Anggraeni & Ratcliffe, 1991)	After separation (Mead et al., 1986)
Plasmatocytes	58.7 ± 3.0	94.5 ± 1.4[c]	67.0 ± 3.6
Granular cells	35.4 ± 2.2	88.5 ± 4.4[c]	94.0 ± 1.4
Spherule cells	4.0 ± 1.9	68.8 ± 12.3[c]	ca.70[d]

[a]Table modified from Anggraeni and Ratcliffe (1991).
[b]Mean value of 12 insects ± S.D.
[c]Mean value from 18 samples from six gradients ± S.D.
[d]Often recovery too low for accurate assessment.

destabilize the cells. Farkǎs (1991), however, reported 87-91% purity for *Galleria* plasmatocytes separated on Ficoll-Paque following prior injection of larvae with the anticoagulant.

CHOICE OF SEPARATION TECHNIQUE

These are many methods available for purifying blood cells, including density gradient centrifugation, velocity sedimentation, centrifugal elutriation, affinity chromatographic, electrophoretic and cellular immunoabsorbent techniques, phase partitioning, and cell sorting (Ali, 1986). Most of these techniques, except density gradient centrifugation and phase partitioning, are unsuitable for insect haemocytes as they require relatively large volumes of blood in contact with columns, gels and chambers which would probably lyse many of the unstable haemocyte types. Undoubtedly, the most promising techniques involve gradient centrifugation in inert polymers of sucrose, such as Ficoll-Paque (Farkǎs, 1991) or non-toxic colloidal silica solutions, such as Percoll (Mead *et al.*, 1986; Anggraeni and Ratcliffe, 1991). Both these substances have low osmotic pressures and allow the separation of cells in anticoagulant-saline solutions with a minimum of manipulation of the fragile cells. Details of the use of Ficoll-Paque have already been given and its disadvantages/ advangates are discussed below.

With Percoll, continuous or discontinuous gradients can easily be formed and both types of gradients should be tested with every new insect species to be used. Continuous gradients have, however, proven their worth with several insect species. A range of continuous gradients from *ca* 50-80% Percoll should be tried in order to generate gradients with different density profiles and to take maximal advantage of the density differences between the various cell types (see Pharmacia Instruction Manual). The gradients are made isotonic with the haemolymph using a specially formulated anticoagulant saline solution. Typically, this solution consists of a ten times concentrated anticoagulant composed, for *G. mellonella*, of 0.613 M NaOH, 1.903 M NaCl, 0.163 M EDTA (free acid) and 0.103 M citric acid at pH 5.5 (heated at 70°C to dissolve the salts) (Mead *et al.*, 1986). This anticoagulant has a higher pH than that used for bleeding (see above) in order to avoid precipitation of the Percoll at low pH. Preformed 60% gradients for separating the *Galleria* haemocytes are made in 10 ml endotoxin-free, round-bottomed, clear polycarbonate tubes by adding 0.7 ml of the × 10 concentrated anticoagulant (detailed above) to 3.78 ml pure stock Percoll plus 2.52 ml of endotoxin-free water. Continuous gradients are then formed by centrifugation at 22,000 g for 20 minutes at 2°C in a 30° angle head on an MSE, high-speed 18 centrifuge. These gradients can be stored only for 1-2 hours before use due to precipitation of the Percoll in the low pH. To each gradient, *ca* 0.8 ml of the haemolymph/anticoagulant mixture from the bled insects is carefully layered on top and the tube then centrifuged

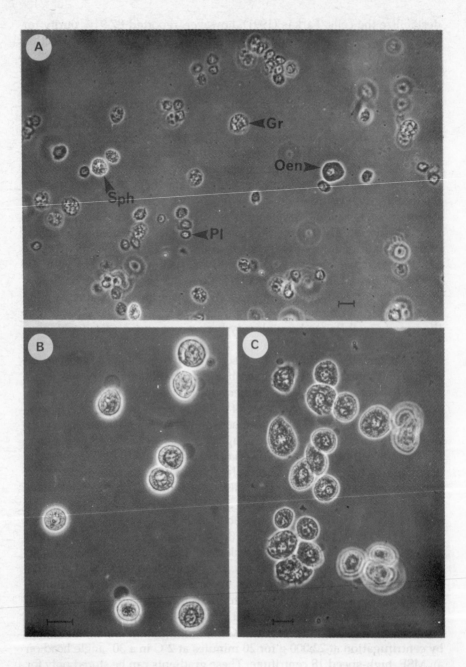

Fig. 1: Haemocytes of *Galleria mellonella*.

A. Haemocytes before separation bled directly onto a slide.
B. Plasmatocytes in anticoagulant following separation on a continuous 60% Percoll gradient.
C. Purified granular cells in anticoagulant following separation on a continuous 60% Percoll gradient.

Oen = Oenocytoid; Pl = Plasmatocyte; Gr = Granular Cell; Sph = Spherule Cell; bars = 10 μm

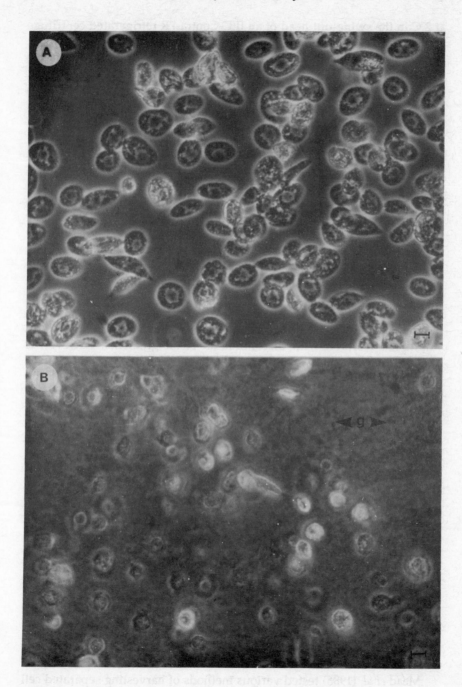

Fig. 2: Haemocytes of *Blaberus craniifer*.

A. Blood cells in anticoagulant; note lack of clotting.

B. Blood cells bled without anticoagulant; note extracellular gelation (g). bars=10 μm

at 2°C in the swing-out head of an IEC-Centra-7R refrigerated centrifuge at 750 g for 20 minutes. No doubt, different combinations of centrifuges and heads would be equally effective at gradient formation and separating the cells. The separated cells form discrete bands which can be harvested with sterile Pasteur pipettes. In the case of *Galleria*, of *ca* 5×10^6 cells loaded on a typical 60% gradient, *ca* 1.0×10^6, 0.65×10^6 and 1.7×10^4 are highly enriched plasmatocytes, granular cells and spherule cells respectively (Fig. 1) (Anggraeni and Ratcliffe, 1991).

The purity of the *Galleria* plasmatocytes and granular cells obtained in the continuous Percoll gradients (Anggraeni and Ratcliffe, 1991) or in the discontinuous Ficoll-Paque gradients (Farkăs, 1991) is comparable and between 87-96% for both cell types. The advantages of the continuous Percoll gradients over the discontinuous Ficoll method are: (1) not only plasmatocytes and granular cells can be separated, but also spherule cells although not as efficiently as in discontinuous Percoll gradients (Cook *et al.*, 1985); (2) continuous Percoll gradients have already been effectively used to separate the blood cells of several different insect species so that the technique is widely applicable; (3) cells harvested from the gradients not only have high viability, but also have been shown to function immunologically *in vitro* (see below) (Anggraeni and Ratcliffe, 1991). Additional research with discontinuous Ficoll gradients, however, is required to assess the versatility of the method and the immunological capability of the cells *in vitro*. Provided the latter are satisfied, then Ficoll gradients do have certain advantages in that: (1) Ficoll-Paque, unlike Percoll is available already made up; (2) the anticoagulant does not, in contrast to Percoll, precipitate Ficoll-Paque; (3) the separation technique is much quicker and therefore less likely to stress the cells; (4) the plasmatocytes and granular cells form more distinct bands than with Percoll (Farkăs, 1991).

HARVESTING CELLS AND SUBSEQUENT EXPERIMENTATION

Despite successful separation of the blood cells from several insect species by Mead *et al.* (1986), these authors failed to obtain sufficient cell attachment to produce monolayers for functional studies. The only such studies, apart from those by Anggraeni and Ratcliffe (1991) (described below), which have been undertaken with separated cells are those of Huxham and Lackie (1988) and Farkăs (1991) and both are very limited in nature. Huxham and Lackie (1988) reported enhanced chemokinesis of locust band 5 cells *in vitro* following activation of prophenoloxidase by β 1-3-glucans, whilst Farkăs (1991) described the *de novo* synthesis of polypeptides by purified populations of plasmatocytes and granular cells from *Galleria*.

Mead *et al.* (1986) tested various methods of harvesting separated cell bands from continuous Percoll gradients and concluded that the most effective technique was with sterile, siliconized Pasteur pipettes from gradients formed in silicon-coated centrifuge tubes.

Another problem relates to the necessity of washing and recentrifuging the harvested cells to remove the anticoagulant and gradient material in order to concentrate the separated cells for *in vitro* work. Recently, Anggraeni and Ratcliffe (1991) examined this problem in detail with separated populations of *Galleria* haemocytes. They set up one group of haemocytes monolayers with cells taken directly from Percoll gradients in the presence of various calcium ion concentrations and showed that maximal attachment (*ca* 29%) to the cover-slips occurred with 40 mM Ca^{2+} in Tris buffer. In the second group of monolayers, the haemocytes were first washed in the anticoagulant, followed by Tris, before resuspension and placement on cover-slips for attachment to occur. Surprisingly, with this latter method, twice as many plasmatocytes attached to the cover-slip than with cells taken directly from the gradients. Subsequent phagocytosis experiments with both types of monolayers showed that only in monolayers formed from washed cells was there any significant ingestion of bacteria. Possibly, the high levels of extraneous calcium, added to the monolayers made with cells taken directly from the gradients, inhibited the prophenoloxidase system (Söderhäll, 1981) and hence the non-self recognition potential of the haemocytes.

Finally, experiments with monolayers made from separated and washed plasmatocytes and granular cells have shown that in *Galleria* the plasmatocyte is the main phagocytic cell type. In addition, adding pure granular cells back into the separated plasmatocytes increased phagocytosis of bacteria from 13.4% to 30.1% (Table 2). Thus, the granular cells, although

Table 2: Percentage of *Galleria mellonella* plasmatocytes and granular cells, alone or mixed, from washed cell preparations phagocytosing *Bacillus cereus* after one hour incubation at 25°C[a]

Type of monolayer	Cell type counted	% of cells phagocytic[b]	
		Control[c]	Experimental[d]
Plasmatocytes alone	Plasmatocytes	5.3 ± 1.3	13.4 ± 2.1[*]
Granular cells alone	Granular cells	0	0
Mixed	Plasmatocytes	8.7 ± 2.7	30.1 ± 3.5[*†]
	Granular cells	0	0

[a]Table modified from Aggraeni and Ratcliffe (1991).

[b]Mean value of 18 monolayers from six gradients ± S.D.

[c]Monolayers were overlaid with *B. cereus* suspended in Tris overlay buffer.

[d]Monolayers were overlaid with *B. cereus* suspended in Tris overlay buffer containing 0.1 mg ml^{-1} β 1,3-glucan.

[*]Significantly different compared with control (P < 0.05).

[†]Significantly different compared with plasmatocytes alone (P < 0.05).

non-phagocytic by themselves, contain some factor(s) essential for optimal functioning of the plasmatocytes during phagocytosis. We are now in a position with these pure, functioning, haemocyte monolayers to examine the processes of immunorecognition in insects in more detail. There is, however, still much scope for improvement in haemocyte separation techniques before *in vitro* studies of immune reactivity can be optimized for a range of insect species.

ACKNOWLEDGEMENTS

I am grateful to Dr. R. Farkǎs, Institute of Experimental Biology and Ecology, Slovak Academy of Sciences for permission to use unpublished material and to Mr. Ian Tew for his technical support. I am also pleased to acknowledge a grant from the Science and Engineering Research Council (Grant Number GR/F 17421) for part of this work.

REFERENCES

Ali, F.M.K. 1986. *Separation of Human Blood and Bone Marrow Cells*. IOP Publishing Ltd., Bristol.

Anggraeni, T. and N.A. Ratcliffe. 1991. Studies on cell-cell cooperation during phagocytosis by purified haemocyte populations of the wax moth, *Galleria mellonella*. *J. Insect Physiol.* 37: 453-460.

Bohn, H. 1977. Differential adhesion of the haemocytes of *Leucophaea maderae* (Blattaria) to a glass surface. *J. Insect Physiol.* 23: 185-194.

Brillouet, C., M. Leclerc, R.A. Binaghi and G. Luguet. 1984. Specific immune response in the sea star *Asterias rubens*: Production of 'antibody like' factors. *Cell Immunol.* 84: 138-144.

Chain, B.M. and R.S. Anderson. 1982. Selective depletion of the plasmatocytes in *Galleria mellonella* following injection of bacteria. *J. Insect Physiol.* 28: 377-384.

Cook, D., D.B. Stoltz and C. Pauley. 1985. Purification and preliminary characterization of insect spherulocytes. *Insect Biochem.* 15: 419-426.

Frakǎs, R. 1991. Separation of plasmatocytes and granulocytes of the greater wax moth, *Galleria mellonella*, by centrifugation on Ficoll-Paque, and patterns of protein synthesis. *Dev. Comp. Immunol.* (in press).

Gupta, S. and R.A. Good. 1980. Markers of human lymphocyte subpopulations in primary immunodeficiency and lymphoproliferative disorders. *Sem. Hematol.* 17: 1-29.

Huxham, I.M. and A.M. Lackie. 1988. Behaviour *in vitro* of separated haemocytes from the locust, *Schistocerca gregaria*. *Cell Tissue Res.* 251: 677-684.

Johansson, M.W. and K. Söderhäll. 1989. Cellular immunity in crustaceans and the propo system. *Parasitology Today.* 5: 183-190.

Kobayashi, M., M.W. Johansson and K. Söderhäll. 1990. The 76kD cell-adhesion factor from crayfish haemocytes promotes encapsulation *in vitro*. *Cell Tissue Res.* 260: 13-18.

Leonard, C.M., K. Söderhäll and N.A. Ratcliffe. 1985. Studies on prophenoloxidase and protease activity of *Blaberus craniifer* haemocytes. *Insect Biochem.* 15: 803-810.

Mead, G.P., N.A. Ratcliffe and L.R. Renwrantz. 1986. The separation of insect haemocyte types on Percoll gradients: Methodology and problems. *J. Insect Physiol.* 32: 167-177.

Peake, P.W. 1979. Isolation and characterization of the haemocytes of *Calliphora vicina* on density gradients of Ficoll. *J. Insect Physiol.* 25: 795-803.

Price, C.D. and N.A. Ratcliffe. 1974. A reappraisal of insect haemocyte classification by the examination of blood from fifteen insect orders. *Z. Zellforsch. Mikrosk. Anat.* 147: 537-549.

Ratcliffe, N.A., C.M. Leonard and A.F. Rowley. 1984. Prophenoloxidase activation: Non-self recognition and cell cooperation in insect immunity. *Science* 226: 557-559.

Ratcliffe, N.A., G.P. Mead and L.R. Renwrantz. 1986. Insect haemocyte separation—an essential prerequisite to progress in understanding insect cellular immunity. In: *Immunity in Invertebrates*, pp. 3-11 (M. Brehelin, ed.). Springer-Verlag, Berlin.

Ratcliffe, N.A., J.L. Broohman and A.F. Rowley. 1991. Activation of the prophenoloxidase cascade and initiation of nodule formation in locusts by bacterial lipopolysaccharides. *Dev. Comp. Immunol.* 15: 33-39.

Ratcliffe, N.A. and S.J. Gagen. 1977. Studies on the *in vivo* cellular reactions of insects: An ultrastructural analysis of nodule formation in *Galleria mellonella*. *Tissue Cell* 9: 73-85.

Söderhäll, K. 1981. Fungal cell wall β 1,3-glucans induce clotting and phenoloxidase attachment to foreign surfaces of crayfish haemocyte lysate. *Dev. Comp. Immunol.* 5: 565-575.

Söderhäll, K. and V.J. Smith. 1983. Separation of the haemocyte populations of *Carcinus maenas* and other marine decapods, and prophenoloxidase distribution. *Dev. Comp. Immunol.* 7: 229-239.

Söderhäll, K. and V.J. Smith. 1986. Prophenoloxidase-activating cascade as a recognition and defense system in arthropods. In: *Hemocytic and Humoral Immunity in Arthropods*, pp. 251-285 (A. Gupta, ed.). John Willey and Sons, New York.

Söderhäll, K., V.J. Smith and M.W. Johansson. 1986. Exocytosis and uptake of bacteria by isolated haemocyte population of two crustaceans: Evidence for cellular cooperation in the defence reaction of arthropods. *Cell Tissue Res.* 245: 43-49.

CHAPTER 4

Cell-mediated Defence Reactions in Insects

J. P. N. Pathak

Introduction

Compared to vertebrates, the immune system in insects is simple as they lack an antigen-antibody complex, although they are capable of responding very effectively against foreign invaders. In the cellular defence mechanism, haemocytes are able to recognize isografts (self) and allografts (non-self) (Lackie, 1986). Haemocytes react in three different ways to foreign invaders, viz., phagocytosis, nodule formation and encapsulation. These processes differ from each other primarily in the relative size of the foreign material to the responding cell and the extent of the reaction (Gupta, 1985). Phagocytosis is the specialized defence mechanism of a metazoan, which is confined to its specialized cells. In this process haemocytes behave selectively with respect to the objects around them. In insects, certain particles cause the reaction of phagocytosis while others do not; as yet, there is no definite postulation to explain how phagocytosis is induced (Götz and Boman, 1985). Nodule formation by the circulating haemocytes is the reaction against a high dose of particulate material, such as bacteria, fungi, protozoa, or suspension of living and non-living particles (Götz, 1986). On random contact with foreign particles the granular cells discharge a flocculent substance that surrounds the foreign object. The cells aggregate around the entrapped micro-organisms to form a compact capsule or nodule which may or may not undergo melanization. Such isolated foreign bodies may be further encapsulated by plasmatocytes to complete cellular encapsulation. Thus encapsulation is a common response to foreign invaders larger than the haemocytes involved in the immune response and involves enclosing the object in several layers of cells (Vinson, 1990).

The role of haemocytes in the defence mechanism of arthropods has been reviewed by several workers (Ratcliffe and Rowley, 1979; Rowley and Ratcliffe, 1981; Ratcliffe, 1982; Götz, 1986). The present paper reviews the cell-mediated defence mechanism in insects.

Phagocytosis

Phagocytosis was the first of the host defence reactions of animals to be studied. Metchnikoff's (1892) classic work on *Daphnia* sp. and on other arthropods, including insects, is considered the beginning of cellular immunity studies in invertebrates. After him a number of investigations were undertaken *in vivo* and *in vitro* to explain phagocytosis. Most of the work on phagocytosis in insects has been carried out *in vivo*. Most workers have shown that if test particles such as carmine, polysterine beads, India ink, erythrocytes, fungi spores, yeast cell wall, protozoa etc. are injected in the haemocoel, the process of phagocytosis is observed within one hour after infection and most of the objects are removed from circulation by the haemocytes (Jones, 1962; Salt, 1970; Arnold, 1974; Whitecomb *et al.*, 1974). Many early attempts to effect phagocytosis *in vitro* failed (Akesson, 1954; Jones, 1956; Whitten, 1964; Lea and Gilbert, 1966; Scott, 1971) although in some of these studies, ingestion may have occurred but could not be recognized due to the inherent difficulties in identifying intracellular test particles (Smith and Ratcliffe, 1978). More success was obtained with cell cultures (Grace, 1962, 1971; Vago, 1964; Vago and Vey, 1970; Vago, 1972; Landureau *et al.*, 1972). Rabinovitch and Stefano (1970) described a simple monolayer technique for examining phagocytosis recognition by *Galleria mellonella* haemocytes. Similarly, Anderson *et al.* (1973) and Anderson in a series of papers (1974, 1975) have provided much information about the role of haemocytes in the phagocytic activity of bacteria by means of suspension culture and simple monolayer techniques. Ratcliffe and Rowley (1974) studied phagocytosis in a new suspension culture system and noted that the plasmatocytes of *Galleria*, *Calliphora* and *Periplaneta* were the most active cell types in phagocytosis of latex, chick erythrocytes and certain bacteria while the granular cells of *Galleria* also phagocytose particles to a limited extent. Sharma *et al.* (1986) injected the yeast cells in *Poecilocerus-pictus* (Fab.) and noted that yeast cells have modified the morphology of haemocytes. The predominant cell involved in phagocytosis was the plasmatocyte, followed by granular haemocyte.

In cell-mediated defence reactions in insects, two aspects are indispensable—contact between foreign body and haemocytes and the cell types involved.

Contact between Foreign Body and Haemocytes

The first step in phagocytosis or any cell-mediated defence reaction is contact between the 'foreign body' and the haemocyte, surface (Ratcliffe and Rowley, 1979). Two concepts prevail concerning this reaction. According to one, a number of authors have tried to find evidence for chemotactic attraction caused by foreign materials (Metchnikoff, 1901; Metalnikov, 1927; Vey *et al.*, 1968; Vey and Vago, 1971). Most of the evidence to indicate chemotactic behaviour of insect haemocytes was gathered from experiments on the encapsulation process and not on phagocytosis. For example, Nappi and Stoffolano (1972) studied the encapsulation of nematodes in *Musca domestica* and *Musca automnalis* larvae and provided circumstancial evidence in favour of the chemotactic factor which helps in premature movement of haemocytes to encapsulate the parasite. However, Ratcliffe *et al.* (1976) have reported more direct evidence for the presence of haemocyte chemotaxis in insects. Similarly, the haemocytes of *G. mellonella* appeared to be attributable *in vitro* to *Aspergillus flavus* conidia (Vey, 1969; Vey and Vago, 1971). Jones (1956) failed to observe any movement of *Sarcophaga bullata* haemocytes towards foreign agents. Salt (1970) also believed that chemotaxis was probably not involved in recognition of foreignness. His conclusions were based on encapsulation of intraspecific organ transplants and inert objects. It was observed that in intraspecific organ implants, only injured parts were sealed by haemocytes. Salt (1970) presumed that if the haemocytes had responded to a chemotactic stimulus, these cells would also have attached to the non-injured part of the implanted tissue. However, Nappi (1974) believed that localized aggregation of the haemocytes on the damaged tissues could be the specific movement of cells to the response received from the 'wound'. Such specific mobilization of haemocytes has also been observed by several other workers (Wigglesworth, 1937; Harvey and Williams, 1961; Lea and Gilbert, 1961; Cherbas, 1973). Salt (1970) further strengthened his concept by stating that a piece of glass or polyfluorocarbon is an inert particle and chemotactic substances would not diffuse from them. He concluded that possibly the cells respond to a substance produced by interaction between a foreign substance and a specific population of haemocytes. Lackie (1986) believed that insect haemocytes are able to recognise self, non-self and damaged self-surfaces. Any alternation in the tissue surface, whether by mutation, or by physical or chemical damage will provoke a haemocytic 'wound-healing' response during which haemocytes aggregate around the altered regions.

Cell Types

The plasmatocytes and granular haemocytes are the predominant cells involved in phagocytosis in various insects. A number of workers have reported the active involvement of plasmatocytes in phagocytosis *in vivo* (Wittig, 1965; Salt, 1970; Ratcliffe and Rowley, 1975). However, due to confusion in the terminology of insect haemocytes these cells have been variously named as leucocytes, lymphocytes, spherule cells, plasmatocytes, adipohaemocytes, micro- and macroplasmatocytes etc. by different workers. In insects, the principal haemocyte with phagocytic activity is the plasmatocyte (Jones, 1975). In *Bombyx mori*, Wago and Ichikawa (1979) studied the changes in phagocytic rate during larval development and the manner whereby haemocytic reactions to foreign cells are initiated. They noted that the ability of haemocytes to contact with GRBCs at the final instar was about seven times as rapid at the 1st instar. They further concluded that the granular cells involved in phagocytosis were of a large type and plasmatocytes were not involved in this process. Wago (1983) described phagocytosis by the granular cells and divided the process into three phases—attachment, filopodial elongation and internalization by the active extension of the veil-like membrane processed. He suggested that filopodial elongation plays a key role in triggering or regulating cellular reactions by the granular cells, particularly in determining either phagocytosis or early phase of encapsulation reaction. However, Pathak (1991) also studied phagocytosis in *B. mori* using three different objects, viz. (bacterium *B. subtiles*, spores of *Aspergillus* and sheep erythrocytes) and noted that *in vitro* bacteria were engulfed by plasmatocytes as well as by granular cells (Fig. 1). The spores of *Aspergillus niger* were engulfed by only granular cells *in vivo* and *in vitro* (Fig. 2). Sometimes the spores were encircled by several degenerating granular cells (Fig. 3). The SRBCs formed rosettes with the granular cells initially and ultimately were engulfed and surrounded by several degenerating granular cells.

Although the cells involved in the process of phagocytosis in insects are well recognized, the mechanism or the factors involved in this process and the role they play are not yet clearly understood. Recently, considerable information has been garnered about the nature of the surface receptors and the recognition of foreignness by insect haemocytes. McSweegan and Pistole (1982) have also provided some circumstantial evidence that serum lectins may act as mediators in specific recognition in arthropods. Induction of such factors by injury in the flesh fly, *Sarcophaga peregrina* has been demonstrated by Komano *et al.* (1981, 1983). They concluded that the galactose-binding lectin developed due to injury is responsible for recognition of disintegrated tissue fragments. They further confirmed that such lectins also develop during metamorphosis (see Chapter 11, Haemagglutinins in Insects).

Nodule Formation

Nodule formation is a common phenomenon in response to both animate and inanimate substances that cannot be removed from circulation by phagocytosis (Metchnikoff, 1884; Bücher, 1959; Salt, 1970). In the process of nodule formation, a coagulum is produced by the degenerating granular cells which is centrally melanized. Particulate and non-particulate substances are entrapped in that coagulum in the centre and surrounded by a sheath of blood cells (Hollande, 1930; Vey, 1968; 1969; Vey *et al.*,

1968, 1973; Vey and Vago, 1971; Gagen and Ratcliffe, 1976; Ratcliffe and Gagen, 1976; Ratcliffe and Rowley, 1979).

A detailed analysis of nodule formation has been provided by Gagen and Ratcliffe (1976) and Ratcliffe and Gagen (1977) working with *Galleria mellonella*. They injected live or dead *Bacillus cereus, E. coli, Sarcina lutea* and *Staphylococcus aurens* into the haemocoel of 5th instar larvae of *G. mellonella* and *P. brassicae*, then killed and dissected the larvae at regular intervals to study nodule formation. These workers recorded that nodules had formed within 1-5 minutes in all the insects except those injected with *S. aurens*. They noted that in *G. mellonella*, the first stage involved the degeneration of granular cells to discharge their granules to entrap the micro-organisms. They further noted that in the next one hr. melanization of degenerated haemocytes and entrapped bacteria had taken place and clumped material had changed into a melanotic mass of necrotic cells (Ratcliffe and Rowley, 1979). In the next 1 to 6 hours, the second stage of nodule formation took place and several layers of plasmatocytes encircled the central melanotic core.

Pathak (unpublished data) observed that in naturally infected last instar larvae of *B. mori*, the clumped masses of bacteria and degenerated granular cells circulated in the haemocoel (Fig. 4). This mass was neither melanized nor encircled by the plasmatocytes (Fig. 5). However, the formation of normal nodules was also noted (Fig. 6).

Encapsulation

Encapsulation of large foreign bodies is a common phenomenon of cell-mediated immune reactions in insects. The foreign invaders are enclosed in several layers of cells and the capsule so formed melanized and isolated from active circulation to protect the host. A number of reviews have been published to explain this cellular defence reaction of insects (Shapiro, 1969; Salt, 1975; Poinar, 1974; Ratcliffe and Rowley, 1979; Rowley and Ratcliffe, 1981; Ratcliffe, 1982; Götz and Boman, 1985; Götz, 1986). These studies discuss the concept of capsule formation and the factors involved in the process of encapsulation.

The literature also provides information about capsule formation on introducing such objects as Araldite (Grimstone *et al.*, 1967; Reik, 1968; Francois, 1975; Hillen, 1977), cellophane and glass rods (Matz, 1965; Brehélin, *et al.*, 1975; Zachary *et al.*, 1975; Salt, 1970), latex (Lackie, 1976), nylon fibres (Hillen, 1977; Sato *et al.*, 1976), biological implants (Lackie, 1976; Smith and Ratcliffe, 1978), nematodes (Poinar *et al.*, 1968; Poinar and Leutenegger, 1971; Nappi and Stoffolano, 1971, 1972; Nappi, 1974, 1975), insect parasitoids (Salt, 1963, 1970, 1973, 1975; Van den Bosch, 1964; Vinson, 1972, 1977), protozoa and fungi (Salt, 1970; Götz and Vey, 1974). Pathak (1991) has given details of encapsulation in the last larval stage of *B. mori*. He noted that granular cells adhere to the foreign object within 3 minutes

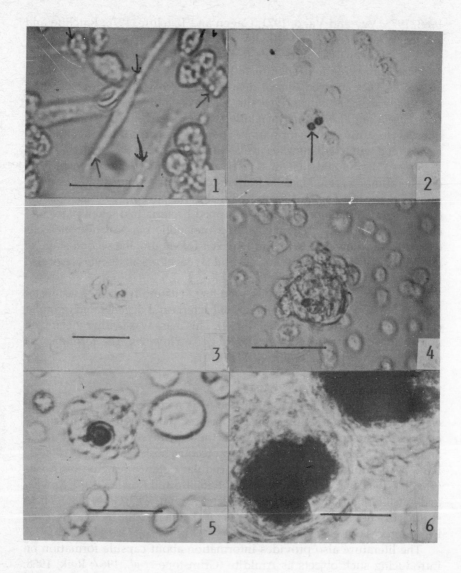

Fig. 1: Phagocytosis of bacteria by plasmatocytes and granular cells. bar = 5 μm.
Fig. 2: Phagocytosis of *Aspergillus* spores by granular cells. bar = 7 μm.
Fig. 3: Aspergillus spores encircled by degenerating granular cells. bar = 7 μm.
Fig. 4: Clumped mass of bacteria and degenerated granular cells. bar = 10 μm.
Fig. 5: Nodule without plasmatocytes. bar = 10 μm.
Fig. 6: Nodule formation by granular cells and plasmatocytes. bar = 5 μm.

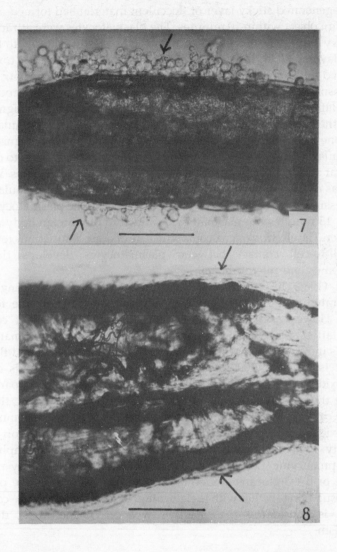

Fig. 7: Random contact of granular cells of *B. mori* around the nerve cord. bar = 5 µm.
Fig. 8: Encapsulation of foreign nerve cord by the haemocytes of *B. mori*. bar = 5 µm.

and a degenerated sticky layer of flocculent material had formed around the foreign object within 10 minutes. The plasmatocytes were attracted by this sticky substance and flattened over the flocculent material to form a several-layered thick capsule (Figs. 7 and 8). The whole structure had melanized within one hour in *in vitro* studies. Gupta (1985), Götz (1986) and Vinson (1990) described the time-bound steps in the process of encapsulation. According to these authors, when a large foreign body reaches the haemocoel, it randomly comes into contact with granular cells within one minute. In the second step, the foreign body is recognized by the granular cells as a non-self structure and the cells degenerate to release a granular sticky substance around the foreign object in 5 minutes. Vinson (1990) has described the flocculent material released by the granular cells as possessing some recognisable factor that attracts the plasmatocytes. In the next 15 minutes the plasmatocytes attach to the foreign body and to each other. The electron microscopic structure of these associations revealed that cell-to-cell contact between plasmatocytes develops through desmosomes and microtubules.

Götz (1986) has mentioned that melanosis occurs, starting from disintegrated granular cells and discharged material near the foreign surface. However, melanization is not a regular process of cellular encapsulation. Only living organisms, tissue and certain inanimate materials induce both encapsulation and melanization (Götz and Boman, 1985). It is believed that melanization of the capsule depends on the phenoloxidase system, which kills the micro-organisms and parasites, isolating the foreign objects from the rest of the body by sealing them in a hard capsule of melanin (Taylor, 1969). Therefore, melanin and phenoloxidase are considered the important components of insect immunity. Phenoloxidase is normally present in the haemolymph in an inactive proenzyme (Leonard *et al.*, 1985) which can be activated by a serine protease or β 1-3-glucans (Söderhäll and Smith, 1986). Thus the process of encapsulation together with melanization by the phenoloxidase cascade reaction is considered the effective weapon of the insect defence mechanism.

Conclusion

Insects lack immunoglobulins such as recognition molecules. However, they possess reactive haemocytes which are capable of responding to a high degree of specificity against a diversity of foreign components. Using such host defense reactions as phagocytosis, nodule formation, encapsulation and agglutination, insects have managed to defend themselves against foreign invaders. The plasmatocytes and granular cells are the main phagocytic cells while the degenerated granular cells and plasmatocytes are responsible for nodule formation and encapsulation.

REFERENCES

Akesson, B. 1954. Observations on the haemocytes during the metamorphosis of *Calliphora erythrocephala* (Meig). *Ark. Zool.* 6: 203-211.

Anderson, R.S. 1974. Metabolism of insect haemocytes during phagocytosis. In: *Contemporary Topics in Immunology. Invertebrate Immunology*, pp. 47-54 (E.L. Cooper, ed.). Plenum Press, New York.

Anderson, R.S. 1975. Phagocytosis by invertebrate cells *in vitro*: Biochemical events and other characteristics compared with vertebrate phagocytic systems. In: *Invertebrate Immunity*. pp. 152-180 (K. Maramorosch and R.E. Shope, eds.). Academic Press, New York.

Anderson, R.S., B. Holmes and R.A. Good. 1973. Comparative biochemistry of phagocytizing insect haemocytes. *Comp. Biochem. Physiol.* 46B: 595-602.

Arnold, J.W. 1974. The haemocytes of insects. In: *The Physiology of Insecta*, vol. 5, pp. 201-255 (M. Rockstein, ed.). Academic Press, New York.

Brehélin, M., J.A. Hoffmann, G. Matz and A. Porte. 1975. Encapsulation of implanted foreign bodies by hemocytes of *Locusta migratoria* and *Melolontha melolontha*. *Cell Tissue Res.* 160: 283-289.

Bücher, G.E. 1959. Bacteria of grasshoppers of western Canada. III. Frequency of occurrence, pathogenicity. *J. Insect Pathol.* 1: 391-405.

Cherbas, L. 1973. The induction of an injury reaction in cultured haemocytes from saturniid pupae. *J. Insect Physiol.* 19: 2011-2023.

Francosis, J. 1975. L' Encapsulation haémocytarie expérimentale chez le lépisme *Thermobia domestica*. *J. Insect Physiol.* 21: 1535-1546.

Gagen, S.J. and N.A. Ratcliffe. 1976. Studies on the *in vivo* cellular reactions and fate of injured bacteria in *Galleria mellonella* and *Pieris brassicae* larvae. *J. Invertebr. Pathol.* 28(1): 17-24.

Götz, P. 1986. Encapsulation in Arthropods. In: *Immunity in Invertebrates*. pp. 153-170. (M. Brehélin, ed.). Springer-Verlag, Berlin-Heidelberg.

Götz, P. and A. Vey. 1974. Humoral encapsulation in Diptera (Insecta) defense reactions of *Chironomous* larvae against fungi. *Parasitology* 68: 1-13.

Götz, P. and H.G. Boman. 1985. Insect Immunity. In: *Commprehensive Insect Physiology, Biochemistry and Pharmocology*. Vol. 3, p. 453 (G.A. Kerkut and L.I. Gilbert, eds.). Pergamon Press, New York.

Götz, P., I. Roettgen and W. Lingg. 1977. Encapsulment humoral en tant que réaction de défense chez les Diptères. *Ann. Parasitol.* 52(1): 95-97.

Grace, T.D.C. 1962. The development of a cytoplasmic polyhydrosis in insect cells grown in *in vitro*. *Virology* 18: 33-42.

Grace, T.D.C. 1971. The morphology and physiology of cultured invertebrate cells. In: *Invertebrate Tissue Culture*, vol. I, pp. 171-209, (C. Vago, ed.). Academic Press, New York.

Grimstone, A.V., S. Rotheram and G. Salt. 1967. An electron-microscope study of capsule formation by insect blood cells. *J. Cell. Sci.* 2: 201-292.

Gupta, A.P. 1985. Cellular elements in the hemolymph. In: *Comparative Biochemistry, Physiology and Pharmacology of Insects*, vol. 3, p. 401 (G.A. Kerkut and L.I. Gilbert, eds.). Pergamon Press, New York.

Harvey, W.R. and C.M. Williams. 1961. The injury metabolism of the cecropia silkworm. I. Biological amplification of the effects of localised injury. *J. Insect Physiol.* 7: 81-99.

Hillen, N.D. 1977. Experimental studies on the reactions of insect haemocytes to artificial implants and habitual parasitoids on the initiation of wound healing in insects. Ph.D. thesis. University of London, London.

Hollande, A.C. 1930. La digestion des bacilies tuberculeux par les leucocytes du sang des chenilles. *Arch. Zool. Exp. Gen.* 70: 231-280.

Jones, J.C. 1956. The haemocytes of *Sarcophaga bullata* Parker. *J. Morphol.* 99: 233-257.

Jones, J.C. 1962. Current concepts concerning insect haemocytes. *Amer. Zool.* 2: 209-246.

Jones, J.C. 1975. Forms and functions of insect hemocytes. In: *Invertebrate Immunity*, pp. 119-128. (K. Maramorosch and R.E. Shope, eds.). Academic Press, New York.

Komano, H., D. Mizuno and S. Natori. 1981. A possible mechanism of induction of insect lectin. *J. Biol. Chem.* 256: 7087-7089.

Komano, H., R. Nozwa, D. Mizuno and S. Natori. 1983. Measurement of *Sarcophaga peregrina* lectin under various physiological conditions by radioimmunoassay. *J. Biol. Chem.* 258: 2143-2147.

Lackie, A.M. 1976. Evasion of the haemocytic defense reaction of insects by larvae of *Hymenolepis diminuta* (Cestoda). *Parasitology* 75: 91-104.

Lackie, A.M. 1986. Transplantation immunology in Arthropods: Is immunorecognition merely wound healing. In: *Immunity in Invertebrates*, pp. 125-136 (M. Brehélin, ed.). Springer-Verlag, Berlin-Heidelberg.

Landureau, J.C., P. Grellet and I. Bernier. 1972. Caractérisation en culture *in vitro* d'un role inconnu des hémocytes d'Insectes: Sa signification physiologique. *C.R.S. Acad. Sci. Paris* 274D: 2200-2203.

Lea, M.S. and L.I. Gilbert. 1961. Cell division in silkworm pupae. *Amer. Zool.* 1: 368-369.

Lea, M.S. and L.I. Gilbert. 1966. The hemocytes of *Hyalophora cercopia* (Lepidoptera). *J. Morphol.* 118(2): 197-216.

Leonard, C.M., K. Söderhäll and N.A. Ratcliffe. 1985. Isolation and properties of the prophenoloxidase activating system in an insect, *Blaberus craniifer*. *Insect Biochem.* 15: 803-810.

Matz, G. 1965. Implantation de fragments de cellophané chez *Locusta migratoria* L. Bull. Soc. Zool. Fr. 90: 429-433.

McSweegan, E.E. and T.G. Pistole. 1982. Interaction of the lectin limulin with capsular polysaccharides. *Biochem. Biophys. Res. Commun.* 106: 1390-1397.

Metalnikov, S. 1927. L' infection microbienne et l'immunité chez la mites des abeilles *Galleria mellonella*. Monograph. Institut Pasteur Masson, Paris.

Metchnikoff, E. 1884. Ueber eine sprosspilzkrandheit der Daphniean: Beitrag zur Lehre uber den Kampf der Phagocyten gegen Krankheitserreger. *Virchow's Arch. Pathol. Anat. Physiol.* 96: 177-95.

Metchnikoff, E. 1892. Lecons sur la pathologie comparee de l'inflammation. Masson, Paris.

Metchnikoff, F. 1901. L'immunite dans les maladies infectieuses. Masson, Paris.

Nappi, A.J. 1974. Insect hemocytes and the problems of host recognition of foreignness. In: *Contemporary Topics in Immunology Invertebrate Immunology*, vol. 4, pp. 201-224 (E.L. Cooper, ed.). Plenum Press, New York.

Nappi, A.J. 1975. Parasite encapsulation in insects. In: *Invertebrate Immunity*, pp. 293-326, (K. Maramorosch and R.E. Shope, eds.). Academic Press, New York.

Nappi, A.J. and J.C. Stoffolano, Jr. 1971. *Heterotylenchus autumnalis*: Hemocytic reactions and capsule formation in host, *Musca domestica*. *Exp. Parasitol.* 29: 116-125.

Nappi, A.J. and J.C. Stoffolano, Jr. 1972. Distribution of haemocytes in larvae of *Musca domestica* and *Musca automnalis* and possible chemotaxis during parasitization. *J. Insect Physiol.* 18: 169-179.

Pathak, J.P.N. 1990. Haemocyte mediated defense mechanism in *Bombyx mori*. *Indian J. Zoolog. Spectrum*, vol. 1, no. 2, 10-14.

Poinar, G.O., Jr. 1974. Insect immunity to parasitic nematodes. In: *Contemporary Topics in Immunology*: vol. 4, pp. 167-178 (E.L. Cooper, ed.). *Invertebrate Immunology*, Plenum Press, New York.

Poinar, G.O., Jr. and R. Leutenegger. 1971. Ultrastructural investigation of the melanization process in *Culex pipens* (Culicidae) in response to a nematode. *J. Ultrastruct. Res.* 36: 149-158.

Poinar, G.O., Jr., R. Leutenegger and P. Gotz. 1968. Ultrastructure of the formation of a melanotic capsule in *Diabrotica* (Coleoptera) in response to parasitic nematode (Mermithiidae). *J. Ultrastruct. Res.* 25: 293-306.

Rabinovitch, M. and M.J. De Stefano. 1970. Interactions of red cells with phagocytes of the wax moth (*Galleria mellonella*) and mouse. *Exp. Cell Res.* 59: 272-282.

Ratcliffe, N.A. 1982. Cellular defense reactions of insects. In: *Immune Reaction to Parasites*, pp. 233-244 (W. Frank, ed.). Fischer, Stuttgart.

Ratcliffe, N.A. and A.F. Rowley. 1974. *In vitro* phagocytosis of bacteria by insect blood cells. *Nature* (Lond.) 252(5482): 391-392.

Ratcliffe, N.A. and A.F. Rowley. 1975. Cellular defense reactions of insect hemocytes *in vitro*. Phagocytosis in a new suspension culture system. *J. Invertebr. Pathol.* 26: 225-233.

Ratcliffe, N.A. and A.F. Rowley. 1979. Role of hemocytes in defense against biological agents. In: *Development, Forms, Functions and Techniques: Insect Hemocytes*, pp. 331-414. (A.P. Gupta, ed.). Cambridge University, Cambridge.

Ratcliffe, N.A. and S.J. Gagen. 1976. Cellular defense reactions of insect hemocytes *in vitro*: Nodule formation and development in *Galleria mellonella* and *Pieris brassicae* larvae. *J. Invertebr. Pathol.* 28(3): 373-382.

Ratcliffe, N.A. and S.J. Gagen. 1977. Studies on the *in vitro* cellular reactions of insects: An ultrastructural analysis of nodule formation in *Galleria mellonella*. *Tissue Cell*. 9(1): 73-85.

Ratcliffe, N.A., S.J. Gagen, A.F. Rowley and A.R. Schmit. 1976. Studies on insect cellular defense mechanisms and aspects of the recognition of foreignness. In: *Proceedings, First International Colloquium on Invertebrate Pathology*, pp. 210-214 (T.A. Angus, P. Faulkner and A. Rosenfield, eds.). Queens University, Kingston, Ontario.

Reik, L. 1968. Contacts between insect blood cells, with special reference to the structure of the capsule formed about parasites, M.Sc. dissertation. University of Cambridge, Cambridge.

Rowley, A.F. and N.A. Ratcliffe. 1981. Insects. In: *Invertebrate Blood Cells*, vol. 2, pp. 421-488 (N.A. Ratcliffe and A.F. Rowley, eds.), Academic Press, New York.

Salt, G. 1959. Experimental studies in insect parasitism. The reaction of a stick insect to an alien parasite. *Proc. R. Soc.* B146: 93-108.

Salt, G. 1963. The defense reactions of insects to metazoan parasites. *Parasitology* 53: 527-642.

Salt, G. 1970. The Cellular Defense Reactions of Insects. Cambridge Monograph in Experimental Biology, No. 16. Cambridge University Press, Cambridge.

Salt, G. 1973. Experimental studies in insect parasitism XVI. The mechanism of the resistance of *Nemeritis* to defence reactions. *Proc. R. Soc. Lond.* 183: 337-350.

Salt, G. 1975. The fate of an internal parasitoid, *Nemeritis canescens*, in a variety of insects. *Trans. R. Entomol. Lond.* 127(2): 141-161.

Sato, S. and H. Akai. 1977. Development of the hemopoietic organs of the silkworm, *Bombyx mori* L. *J. Seric. Sci. Jpn.* 46: 397-403.

Sato, S., H. Akai and H. Sawada. 1976. An ultrastructural study of capsule formation by *Bombyx mori. Annot. Zool. Jpn.* 49(3): 177-188.

Scott, M.T. 1971. Recognition of foreignness in invertebrates. II. *In vitro* studies of cockroach phagocytic haemocytes. *Immunology* 21: 817-827.

Shapiro, M. 1969. Immunity of insect hosts to insect parasites. In: *Immunity to Parasitic Animals*, vol. 1, pp. 211-228 (G.J. Jockson, R. Herman and I. Singer, eds.), Appleton, New York.

Sharma, P.R., K. Tikku and B.P. Saxena. 1986. An electron microscopic study of normal haemocytes of *Poecilocerus pictus* (Fab.) and their response to injected yeast cells. *Insect Sci. Applic.* 7: 85-91.

Smith, V.J. and N.A. Ratcliffe. 1978. Host defense reactions of the shore crab *Caricinus maenas* (L), *in vitro*. *J. Mar. Biol. Assoc. UK*, 58: 367-379.

Söderhäll, K. and V.J. Smith. 1986. The prophenoloxidase activating cascade as a recognition and defense system in arthropods. In: *Humoral and Cellular Immunity in Arthropods*, pp. 251-285 (A.P. Gupta, ed.). Wiley, New York.

Taylor, R.L. 1969. A suggested role for the polyphenol-oxidase system in invertebrate immunity. *J. Invertebr. Pathol.* 14: 427-428.

Vago, C. 1964. Culture de tissue d'invertébrés. Service du Film Recherche Scientifique, Paris.

Vago, C. 1972. Invertebrate cell and organ culture in invertebrate pathology. In: *Invertebrate Tissue Culture*, vol. 2, pp. 245-278. (C. Vago, ed.). Academic Press, New York.

Vago, C. and A. Vey. 1970. Mycoses d'invertébrés. Service du Film Recherche Scientifique, Paris.

Van den Bosch, R. 1964. Encapsulation of the eggs of *Bathyplectes curculionis* (Thomson) (Hymenoptera: Ichneumonidae) in larvae of *Hypera brunneipennis* (Boheman) and *Hypera postica* (Gyllenhall) (Coleoptera: Curculionidae). *J. Insect. Pathol.* 6: 343-367.

Vey, A. 1968. Reactions de defense cellulaire dans les infections de blessures a *Mucor heimalis* Wehmer. *Ann. Epiphyt.* (Paris) 19: 695-702.

Vey, A. 1969. Etude *in vitro* des reactions anticryptogamique des larves de Lepidoptereres: Colloque sur les manifestations inflammatoires et tumorales chez les invertebres, 1967. *Ann. Zool. Ecol. Anim.* 1: 93-100.

Vey, A. and C. Vago. 1971. Reaction anticryptomique de type granulome chez les insects. *Ann. Inst. Pasteur (Paris)* 121: 527-532.

Vey, A., J.M. Quiot and C. Vago. 1968. Formation *in vitro* de reactions d'immunite cellulaire chez les insectes. In: *Proceedings, Second International Colloquium on Invertebrate Tissue Culture*, pp. 254-263. Instituto Lombario di Scienze e Lettere, Milan.

Vey, A., J.M. Quiot and C. Vago. 1975. Mise en evidence et etude l'action d'une mycotoxine, la beauvercine, sur des cellules d'insects cultivees *in vitro*. *C.R. Hebd. S. Acad. Sci. Paris*, Ser. D. 276: 2489-2492.

Vinson, S.B. 1972. Effect of the parasitoid *Campoletis sonorensis* on the growth of its host, *Heliothis virescens*. *J. Insect Physiol.* 18: 1501-1516.

Vinson, S.B. 1977. *Microlitis croceipes*: Inhibitions of the *Heliothis zea* defense reaction to *Cardiochiles nigriceps*. *Exp. Parasitol.* 41: 112-117.

Vinson, S.B. 1990. *Immunosuppression in New Direction in Biological Control: Alternatives for Suppressing Agricultural Pests and Diseases*, pp. 517-535. Alan R. Liss, Inc., USA.

Wago, H. 1983. The important significance of filopodial function of phagocytic granular cells of the silkworm *Bombyx mori* in recognition of foreignness. *Devel. and Comp. Immunol.* 7: 445-453.

Wago, H. and Y. Ichikawa. 1979. Changes in the phagocytic rate during larval development and manner of haemocytic reactions to foreign cells in *Bombyx mori*. *Appl. Ent.* 14(4): 397-403.

Whitcomb, R.F., M. Shapiro and R.R. Granados. 1974. Insect defense mechanisms against microorganisms and parasitoids. In: *The Physiology of Insecta*, vol. 5. (M. Rockstein, ed.). Academic Press, New York, 2nd ed.

Whitten, J.M. 1964. Haemocytes and the metamorphosing tissues in *Sarcophaga bullata*, *Drosophila melanogaster* and other cyclorrhaphous Diptera. *J. Insect Physiol.* 10: 447-469.

Wigglesworth, V.B. 1937. Wound healing in an insect, *Rhodnius prolixus* (Hemiptera). *J. Exp. Biol.* 14: 364-381.

Wittig, G. 1965. Phagocytosis by blood cells in healthy and diseased caterpillars. I. Phagocytosis of *Bacillus thuringiensis* Berliner in *Pseudaletia unipuncta* (Haworth). *J. Invertebr. Pathol.* 7: 474-488.

Zachary, D., M. Brehelin and J.A. Hoffmann. 1975. Role of the 'thrombocytoids' in capsule formation in the dipteran *Calliphora erythrocephala*. *Cell Tissue Res.* 162: 343-348.

CHAPTER 5

Humoral Encapsulation

Alain Vey

Introduction

In some insects the penetration of invaders triggers an encapsulation process consisting of the formation of a melanotic material around the foreign organism without the visible participation of haemocytes. This mechanism is of special interest as it is very efficient against micro-organisms and parasites. It is also a very good model for investigations on the mechanisms intervening in melanization and in the defence reactions of insects.

Main Characteristics of Humoral Encapsulation

The formation of a sheath of melanotic material accumulating on the surface of parasites was first described in mosquitoes and chironomid larvae parasitized by nematodes (Wülker, 1961; Bronskill, 1962; Esslinger, 1962). Later, Götz (1969, 1973) demonstrated that this reaction occurred both *in vivo* and *in vitro* in the haemolymph of chironomid larvae (Figs. 1 and 3).

A comparative study of larvae and adults of 38 species from 12 insect orders (Götz *et al.*, 1977) established that humoral encapsulation occurs only in certain dipteran species of the families of Culicidae, Chironomidae, Psychodidae, Syrphidae and Stratiomyidae. Humoral encapsulation was observed only in dipteran species in which the total haemocyte count (THC) was very low (< 6000 cells/mm³). On the contrary, cellular encapsulation occurred in dipteran species with high THC as well as in the other insect orders, and more generally in arthropods. However, humoral encapsulation has been observed in a non-dipteran insect host, the potato leafhopper, *Empoasca fabae* (Homoptera: Cicadellidae) infected by the fungus *Erynia radicans* (Entomophtorales) (Butt *et al.*, 1988). Studies on the defence reactions of crustaceans have also shown that reactions related to humoral encapsulation may be observed among the decapods, e.g., in the freshwater crayfish attacked by the crayfish plague fungus, *Aphanomyces astaci* (Söderhäll, 1982).

When humoral encapsulation was studied *in vitro* in isolated haemolymph, the deposition of capsular material at the surface of the provocateurs became visible within 2-5 minutes of incubation. This material appeared first as droplets on the provocative surfaces, which increased in number to form a complete cover. Initially the capsular material was soft and sticky. The capsule was thin, flexible and difficult to detect, and the encapsulated nematodes were still able to move. The encapsulated organisms showed an increasing tendency to adhere to glass or any other surface they happened to contact. After a few minutes the capsule hardened and the nematodes enclosed within the rigid envelope. The capsules formed in *in vitro* conditions reached a thickness of about 0.5 to 1.0 μm. After 1-2 hours they changed from yellow to brown (bright orange under phase contrast).

During the *in vitro* experiments, foreign surfaces other than those of the parasites or micro-organisms also provoked the formation of granular material, such as glass surfaces or the zones of contact of the haemolymph with air. This reaction was not so intense as that against bacteria, fungi or nematodes. However, the reaction against these surfaces decreased the intensity of the anti-parasitic or anti-microbial reaction and resulted in about 30 minutes in the complete disappearance of the encapsulation capacity of the sample of isolated haemolymph (Götz *et al.*, 1987).

Electron microscopic studies of humoral encapsulation have enabled observation of the ultrastructure of the capsule material. During the first minutes after injection in the host (2-15 minutes) a loose aggregate of fibrillar material accumulates around the provocative agents. During the next 15-30 minutes this aggregate becomes denser and within a few hours the capsule has reached a thickness of one to several μm (Götz and Vey, 1974). It then shows a high electron density and an irregular fibrillar structure.

Agents Provoking Humoral Encapsulation

Humoral encapsulation can be triggered by stimuli of a very different nature.

NEMATODES

The formation of dark non-cellular incrustations was first observed at the surface of mermithid larvae developing in chironomids (Wülker, 1961). These early studies on encapsulation of parasitic nematodes had already emphasized the rapidity of the reaction (occurring within minutes) following the penetration of the parasite into the haemocoel of the host. In the haemocoel of 'non-typical' host species, all the parasitic nematodes were encapsulated and died. In 'adapted' host species, a certain percentage of the mermithids remained alive and were able to continue development (Götz and Vey, 1986). The natural rate of infection resulting from this percentage survival of the parasites ranged from 0.1 to 10%.

FUNGI

Fungal elements also provoke humoral encapsulation *in vitro* and in the insect host (Figs. 1 and 2). Injection of the conidia of *Beauveria bassiana*, *Mucor hiemalis* and *Aspergillus niger* into *Chironomus* larvae was followed by intense encapsulation of all these fungi (Götz and Vey, 1974). Most of the spores germinated in spite of their encapsulation and the outgrowth of germ tubes provoked an additionnal deposition of melanotic material. The further growth of *A. niger*, a saprophytic fungus, was blocked by the humoral encapsulation. On the contrary, the conidiospores of *B. bassiana*, a highly pathogenic and fast-growing fungus, produced filaments able to break the capsular envelope and to exhaust the encapsulating capacity of the host. Its hyphae invaded the haemocoel of the larvae, which died within 2-5 days at 22°C.

Among the fungi studied, only *B. bassiana* was able to penetrate through the integument of chironomids. The penetrating hyphae provoked the deposition of dark electron-dense material at the surface of the fungal filaments within the cuticle and the hypodermal cells. The defence reaction at these levels was similar to humoral encapsulation observed in the haemocoel after injection. The compounds necessary for the formation of a capsule are therefore present in the cuticle as well as in the haemolymph.

BACTERIA

The haemolymph of *Chironomus* members shows a low cellular and lytic activity against bacteria. Thus phagocytosis occurs late and with a low frequency in injected *Chironomus* larvae. Formation of haemocytic nodules around aggregates of bacterial cells was never observed (Götz *et al.*, 1987). Moreover, humoral antibacterial activity characterized by the synthesis of substances such as lysozyme, cecropins and attacins (Mohrig and Messner, 1968; Faye *et al.*, 1975), was not detected in species of *Chironomus*. Thus larvae 'immunized' by an injection of cells of *Enterobacter cloacae* revealed a very low level of inhibitory activity against *E. coli*.

Contrarily, the chironomid larvae showed an efficient capacity to quickly encapsulate these prokaryotic micro-organisms (Götz *et al.*, 1987). The fate of the bacteria was followed *in vivo* after injection and *in vitro* in isolated haemolymph. A melanotic capsule was formed after an injection of all the gram-positive and negative strains of bacteria tested. Most of the bacterial cells were completely enclosed in the haemocoel of the insect host after 10 minutes. The accumulation of material went on for many hours before the capsules reached a thickness of several μm. The coated bacteria became attached to the tissues bordering the haemocoel, such as the hypodermis and the phagocytic tissue.

Even the species of bacteria highly pathogenic for other insects were harmless for the *Chironomus* larvae, except at doses higher than 10 bacterial cells per larva. Contrarily, the rate of mortality of the larvae of a dipteran insect, *Galleria mellonella,* was correlated to the type of bacteria used and the highly pathogenic species killed 80-100% of the *Galleria* at doses as low as 10 cells/larvae. Thus, humoral encapsulation of bacteria appears to be more efficient than cellular defence due to the difference in kinetics of the two types of encapsulation.

OTHER FOREIGN MATERIALS

A large number of other organisms and foreign bodies have been tested for their capacity to provoke humoral encapsulation (Götz, 1969, 1986a and b; Vey and Götz, 1975; Wilke, 1979; Götz and Vey, 1986).

In vitro and *in vivo* experiments have established that 'strong provocateurs', i.e., provocateurs inducing fast and dense encapsulation in more than 80% of the tests carried out (Götz and Vey, 1986), include living organisms such as protozoa (sporozoa, ciliates), injured homologous tissues, heterologous tissues and some organic polymers, such as cotton, silk and agar, as well as neutral or negatively charged polydextrans (Fig. 3).

Some specific organic compounds (hair, epoxy resin, cationic polydextrans) can be classified as 'weak provocateurs' inducing a reaction in 30-80% cases. Finally, 'non-provocateurs', such as inorganic materials (glass powder, iron powder) and some hydrophobic compounds (nylon, PVC), do not provoke encapsulation.

Nature of the Capsule Material: Humoral Encapsulation and Melanization

Due to their brownish or black appearance, the capsules resulting from humoral encapsulation have always been assumed to be of a melanotic nature (Bronskill, 1962; Götz, 1969; Poinar and Leutenegger, 1971). Various tests have been performed to check the real chemical nature of these capsules: effect of organic and inorganic solvents and of bleaching agents on the capsular material, histochemical tests, enzymatic treatments, incubation with substrates and inhibitors of phenoloxidase (PO) (Maïer, 1973; Götz and Vey, 1974; Vey and Götz, 1975). The results of these assays confirmed the presence of melanin in the capsules. However, it was concluded that the capsular material consisted of a protein polyquinone complex (Götz and Vey, 1974) and not of pure melanin.

The encapsulation process comprises two phases (Wilke, 1979; Götz, 1986a and b). The first step consists of the binding of translucent material, probably activated PO, and other sticky proteins, to the foreign activating surface. This phase is Ca^{++} dependent but tyrosinase independent. The second step consists of solidification and tanning of the capsule material and sclerotization. This step is Ca^{++} independent and consumes tyrosine. It produces the brown color and the electron-dense structure characteristic of well-developed capsules. While newly deposited capsule material exhibits PO activity, during the second phase of encapsulation the phenoloxidase activity stops, probably because the available haemolymph tyrosine is completely exhausted.

As humoral encapsulation takes place within minutes of contact of the haemolymph with the surface of provocateurs, it was hypothesized that pro-PO in insects, such as members of *Chironomus*, occurs freely in the plasma and is immediately activated. This hypothesis was confirmed by research on activation of prophenoloxidase in different insect species.

Three different groups have been distinguished among the insects regarding differences at the level of the pro-PO system:
— Most hemimetabolic insects (except Hemiptera) have a haemolymph which does not develop a PO activity within 30 min after sampling at room temperature.
— A second group, which represents the majority of the holometabolic species, has a pro-PO system spontaneously activated within minutes after bleeding.
— In *Chironomus* and *Chaoborus* larvae, which develop a humoral encapsulation process, a PO activity is present immediately after bleeding and no increase in PO activity occurs during exposure to air.

It is very interesting to observe that the insects showing a capacity for humoral encapsulation are the only ones which exhibit a considerable PO activity in fresh haemolymph.

Effect of Different Factors on Humoral Encapsulation

Parasitism of *C. riparius* by the nematode *Hydromermis contorta* resulted in a weakening and postponement of humoral encapsulation (Vey and Götz, 1975), which may have been the consequence of a reduced level of PO activity in parasitized chironomid larvae compared to the non-parasitized controls (Maïer, 1973).

The experimental studies of Lingg (1976) with *H. contorta* and *C. riparius* demonstrated that the pre-parasitic nematodes became encapsulated while the developing parasitic stages did not trigger the encapsulation reaction.

It has also been observed that all the larvae of *H. rosea* were strongly encapsulated and killed when injected into the haemocoel of *Chironomus* but that the reaction was weaker when the larvae of the same parasite naturally invaded the host by penetrating through the cuticle. If naturally invading nematodes provoke a weaker reaction than injected nematodes, this could be due, according to Götz and Vey (1986), to the fact that during the penetration procedure the parasite can undergo changes in the structure of its surface antigens.

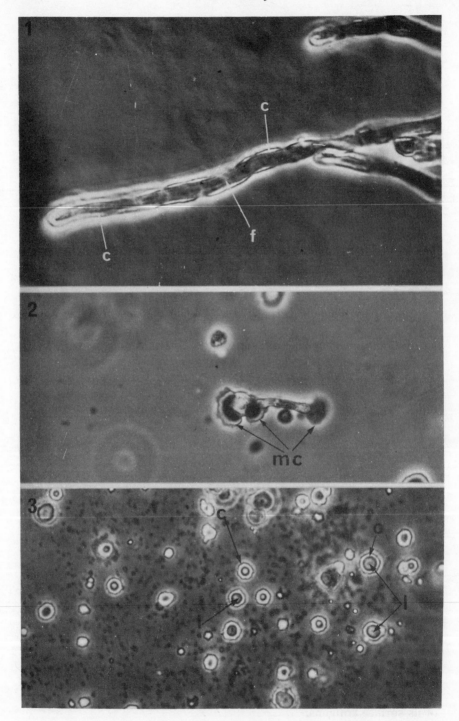

The humoral encapsulation process observed in *in vitro* conditions occurs at different rates depending on the species of insect host investigated. Thus, in *C. luridus* the encapsulation of fungal elements occurred very fast. The capsule was completely formed in 5-6 minutes while in *C. thummi* larvae encapsulation was completed only within 10-15 minutes.

Humoral encapsulation is probably effective in the complete range of temperature to which the chironomid larvae are naturally exposed. However, in *in vitro* conditions low temperatures (4°C) or high temperatures (37°C) provoke some changes in the process. Thus at 4°C the reaction starts at the same time as at 22°C but proceeds more slowly and goes on for a longer time.

Methods of Study

IN VIVO

Humoral encapsulation may be experimentally studied in insect hosts. Suspensions of fungal spores or of bacterial cells can be injected into *Chironomus* larvae. As described by Götz and Vey (1986), the injections are performed by inserting a glass micropipette into one of the prolegs of the larva and closing the wound with a loop of nylon thread tightened around the injected proleg, or by direct perforation of the dorsal abdominal cuticle. The evolution of the encapsulation process is then followed by means of histological or ultrastructural methods.

IN VITRO

Researches in *in vitro* conditions have numerous advantages. Diverse materials can be incubated with isolated haemolymph to comparatively study reactions to different stimuli. The dynamics of the reaction can be observed directly under the microscope and the observation of the process is easy as cells do not participate in the reaction. Furthermore, inhibitors, substrates or lytic enzymes may be added, enabling investigations of the influence of such factors on the humoral encapsulation reaction.

Difficulties are encountered, however, in *in vitro* experiments. Only a few microlitres of haemolymph can be withdrawn from a single larva and a flocculent material appears spontaneously in the sample. This situation

---

Fig. 1: Humoral encapsulation of filaments of the entomogenous fungus *Matarhizium anisopliae in vitro* in a drop of *Chironomus luridus* haemolymph. c—capsule material; f—fungal filament; phase contrast, × 1400

Fig. 2: Formation of capsular deposit at the surface of germinated spores of *Mucor hiemalis* in the haemolymph of *Culex* sp. m.c.—masses of capsular deposit, phase contrast, × 700

Fig. 3: Strong humoral encapsulation of latex particles suspended in *Chironomus luridus* haemolymph. c—capsule; l—latex particles; phase contrast, × 1100

leads to a loss of the encapsulation capacity of the collected haemolymph. It is therefore useful to collect the haemolymph under low temperature and to store it at −20 °C, or at least to use the haemolymph of *Chironomus* immediately after sampling. For microscopic examination of the development of the encapsulation reaction the haemolymph is placed in sterile microchambers in which provocateurs may be added (Vey and Götz, 1975). Simple microchambers can be prepared in every laboratory using a glass slide on which thin strips of vaseline are deposited (in a square of about 1.5 cm of side length). A drop of ringer solution containing organisms or foreign bodies is deposited in the centre of the square. Then a drop of haemolymph is introduced into the chamber which is closed with a cover-slip and sealed by exerting a slight pressure on it. Such preparations can be kept for several days.

Conclusions

Humoral encapsulation is the main defence reaction in certain dipteran species in which the haemolymph produces a melanotic cell-free capsule. This type of encapsulation has been mainly studied *in vivo* and *in vitro* using larvae of *Chironomus* sp.

Cellular and humoral encapsulation have a major characteristic in common: during these two processes the surface of the foreign organism is covered with a melanotic material.

In the humoral encapsulation prophenoloxidase is present in the plasma of the haemolymph and activation of the pro-PO is very rapidly triggered upon contact with provocative agents, which include living organisms and micro-organisms and certain organic polymers. This process is controlled by factors such as β 1-3-glucans (activators of the pro-PO cascade), Ca^{++} and a serine protease.

A comparison of cellular and humoral encapsulation revealed that humoral encapsulation is a fast and very efficient defence reaction mechanism against bacteria, fungi and nematodes, and is even more successful against these invaders than the cellular encapsulation occurring, for example, in lepidopteran larvae.

References

Bronskill, J.F. 1962. Encapsulation of rhabditoid nematodes in mosquitoes. *Can. J. Zool.* 40: 1269-1275.

Butt, T.M., S.P. Wraight, S. Galaini-Wraight, R.A. Humber, D.W. Roberts and R.S. Soper. 1988. Humoral encapsulation of the fungus *Erynia radicans* (Entomophtorales) by the potato leafhopper, *Empoasca fabae* (Homoptera: Cicadellidae). *J. Invert. Pathol.* 52: 49-56.

Esslinger, J.H.Y. 1962. Behaviour of microfilariae of *Brugia pahangi* in *Anopheles quadrimaculatus*. *Am. J. Trop. Med. Hyg.* 1: 749-758.

Faye, I., A. Rasmuson, H.G. Boman and I.A. Boman. 1975. Insect immunity. II. Simultaneous induction of antibacterial activity and selective synthesis of some hemolymph proteins in diapausing pupae of *Hyalophora cecropia* and *Samia cynthia*. *Infect. Immun.* 12: 1426-1438.

Götz, P. 1969. Die Einkapselung von Parasiten in der Hämolymph von *Chironomus* Larven (Diptera). *Zool. Anz.* 33: 610-617 (Supplement).

Götz, P. 1973. Immunreaktionen bei Insekten. *Naturwiss. Rundschau.* 26: 367-375.

Götz, P. 1986a. Encapsulation in arthropods In: *Immunity in Invertebrates*, pp. 153-170 (M. Brehélin, ed.). Springer-Verlag, Berlin-Heidelberg.

Götz, P. 1986b. Mechanisms of encapsulation in dipteran hosts. In: *Immune Mechanisms in Invertebrate Vectors*, (A. Lackie, ed.). Oxford University Press, Oxford.

Götz, P. 1991. Prophenoloxydase activation in haemolymph of different insect species. *Dev. Comp. Immunol.* (in press).

Götz, P. and A. Vey. 1974. Humoral encapsulation in Diptera (Insecta): Defence reactions of *Chironomus* larvae against fungi. *Parasitology* 68: 193-205.

Götz, P. and A. Vey. 1986. Humoral encapsulation in insects. In: *Hemocytic and Humoral Immunity in Arthropods*, pp. 407-429 (A.P. Gupta, ed.). John Wiley and Sons, New York.

Götz, P., I. Roettgen, and W. Lingg. 1977. Encapsulement humoral en tant que réaction de défense chez les diptéres. *Ann. Parasitol. Hum. Comp.* 52: 95-97.

Götz, P., G. Enderlein and I. Roettgen. 1987. Immune reactions of *Chironomus* larvae (Insecta: Diptera) against bacteria, *J. Insect Physiol.* 33: 993-1004.

Lingg, A. 1976. Experimentelle Untersuchungen zur humoralen Einkapselung. Staatsexamensarbeit. University of Freiburg, Germany.

Maier, A.W. 1973. Die Phenoloxydase von *Chironomus thummi* und ihre Beeinflussung durch parasitäre Mermithiden. *J. Insect Physiol.* 19: 85-95.

Mohrig, W. and B. Messner. 1968. Immunreaktionen bei Insekten. I. Lysozym als grundlegender antibakterieller Faktor im humoralen Abwehrmechanismus der Insekten. *Biol. Zbl.* 87: 439-470.

Poinar, G.O. and R. Leutenegger. 1971. Ultrastructural investigations of the melanization process in *Culex pipiens* (Culicidae) in response to a nematode. *J. Utrastruct. Res.* 36: 149-158.

Söderhäll, K. 1982. Prophenoloxidase activating system and melanization—a recognition mechanism of arthropods? A review, *Dev. Comp. Immunol.* 6: 601-611.

Vey, A. and P. Götz. 1975. Humoral encapsulation in Diptera (Insecta): Comparative studies *in vitro*. *Parasitology* 70: 77-86.

Wilke, U. 1979. Humorale Infektabwehr bei *Chironomus* larven. In: *Verhandlugen der Deutschen Zoologischen Gesellschaft*, p. 315 (W. Rathmayer, ed.). Gustav Fischer. Stuttgart, Germany.

Wülker, W. 1961. Untersuchungen über die Intersexualitat der Chironomiden (Dipt.) nach *Paramermis* infektion. *Arch. Hydrobiol.* 25: 127-181.

CHAPTER 6

Inducible Humoral Antibacterial Immunity in Insects

Godwin P. Kaaya

Introduction

Insects have effective immune systems composed of both cellular (Boman et al., 1978; Boman and Hultmark, 1981; Salt, 1970) and humoral components (Boman, 1982; Boman et al., 1986; Boman and Hultmark, 1987; Faye et al., 1975; Götz and Boman, 1985). Cellular immune reactions include phagocytosis, nodule formation and encapsulation (Götz and Boman, 1985; Ratcliffe and Rowley, 1979; Ratcliffe et al., 1985), whereas humoral immune reactions involve synthesis and release of several antibacterial immune proteins (Boman and Hultmark, 1987; Götz and Boman, 1985), some capable of killing both gram-positive and gram-negative bacteria. The expression of this multicomponent humoral immune system requires de novo synthesis of RNA and proteins with broad antibacterial activity. There are at least three families of antibacterial proteins: lysozyme, cecropins and attacins (Boman et al., 1986; Boman and Hultmark, 1987; Boman and Steiner, 1981). Humoral immunity studies in insects have been conducted mostly in lepidopterans (Briggs, 1958; Chadwick, 1967; Hoffmann et al., 1981; Stephens and Marshall, 1962) and in diapausing pupae of silkmoths, e.g. Anthaeraea pernyi, Samia cynthia and Hyalophora cecropia (Boman and Hultmark, 1987; Boman et al., 1974; Faye et al., 1975; Hultmark et al., 1980). Pupae of H. cecropia constitute a particularly good model for studying humoral immunity as they are large (5-10 g) and contain 1-2 ml of haemolymph. They also undergo a long diapause (6-9 months) that allows immune proteins to be synthesized without much interference from other biosynthetic processes (Boman et al., 1986; Boman and Hultmark, 1981).

This paper is concerned exclusively with the inducible humoral antibacterial factors of insects and explains the techniques used to immunize insects for detection of immune proteins and antibacterial activity. Hopefully, it will provide students with the basic knowledge required for conducting experiments on those insect species which have not yet been studied.

Bacteria Used in Assays

The bacteria commonly used in bacteriolysis assyas are dried *Micrococcus luteus* (for lysozyme) from Sigma Chemicals and log-phase *Escherichia coli* K-12, strain D31, an ampicillin- and streptomycin-resistant mutant with a defective lipopolysaccharide. *Enterobacter cloacae*, strain B12, resistant to nalidixic acid, as well as *E. coli* D31 have been used for induction of immunity (Boman *et al.*, 1974; Flyg *et al.*, 1987; Hultmark *et al.*, 1982; Hultmark *et al.*, 1980; Kaaya *et al.*, 1987). Heat-killed *Pseudomonas aeruginosa* has been reported to induce immunity in insects (Chadwick, 1970; Gingsrich, 1964) but we now know that live bacteria are much more effective (Boman *et al.*, 1978; Boman and Steiner, 1981; Kaaya and Darji, 1988).

Insects Immunized

Among the insects reported to have been successfully immunized with bacteria are *Galleria mellonella* and some other lepidopterans (Briggs, 1958; Chadwick, 1967; Hoffmann *et al.*, 1981; Stephens and Marshall, 1962); a hemipteran (Gingsrich, 1964); silkmoths (Boman *et al.*, 1974; Faye *et al.*, 1975; Hultmark *et al.*, 1980; Qu *et al.*, 1982); dipterans, e.g. *Drosophila* (Boman *et al.*, 1972; Flyg *et al.*, 1987; Robertson and Postlethwait, 1986), tsetse flies (Kaaya and Darji, 1988; Kaaya *et al.*, 1987), *Phormia terranovae* (Keppi *et al.*, 1986; Keppi *et al.*, 1989), *Sarcophaga peregrina* (Ando *et al.*, 1987; Okada and Natori, 1983); *Manduca sexta* (Dunn and Drake, 1983; Hurlbert *et al.*, 1985); the darkling beetle (Spies *et al.*, 1986b); locusts (Lambert and Hoffmann, 1985) and *Rhodnius prolixus* (De Azambuja *et al.*, 1986). However, much of the available data on insect immunity have been obtained from diapausing pupae of silkmoths, especially *Hyalophora cecropia*.

Immunization Procedure

Diapausing pupae of *S. cynthia* and *H. cecropia* can be stored in a refrigerator at 8 °C. During experiments they are usually stored at 25 °C, with a relative humidity of 60-80% and a 15.5-hour light: 8.5-hour dark photoperiod. They are vaccinated with viable cells of *E. cloacae* B12 or *E. coli* D31 injected into the thorax; for *S. cynthia* about 5 x 10^5 and for *H. cecropia* 10^6 cells in normal saline using an 'AGLA' micrometer syringe (Wellcome Reagents, London) (Boman *et al.*, 1974; Faye *et al.*, 1975; Hultmark *et al.*, 1980) or an Arnold hand micro-applicator (Burkard Scientific (Sales) Ltd., Rickmansworth, England) fitted with a 1-ml glass syringe and a 31-gauge needle. Control pupae are injected with an equal volume of normal saline. To stop excessive bleeding, the injection wound is quickly sealed with a drop of beeswax melted by heating on a Bunsen burner. The wax hardens soon after being dropped on the insect cuticle and seals off the wound.

Injections can also be done with a small graduated syringe (e.g. Hamilton syringe, Hamilton Bonaduz AG, CH7402 Bonaduz, Switzerland) capable of delivering the desired volume of liquid. Haemolymph samples can be collected from the immunized pupae at different intervals after vaccination by making small incisions on the disinfected (70% ethanol) pupal surface using a sterile scalpal blade. The haemolymph can then be aspirated into a capillary tube and the incision sealed with beeswax. The haemolymph should be transferred into ice-cooled Eppendorf tubes containing a few crystals of phenylthiourea (PTU) to inhibit melanization. It should be tested immediately or stored at $-20\,°C$ until required.

Adult tsetse flies are anaesthetized either by exposure to CO_2 or by chilling in a refrigerator at $4\,°C$ for approximately 1 minute. The insects are disinfected with 70% ethanol and bacteria are injected into their thorax (in 0.5-5 µl saline) using an Arnold hand micro-applicator (Kaaya *et al.*, 1986a and b). This technique has also been used for injecting larvae of the lepidopteran cereal stem borers, *Chilo partellus* and *Busseola fusca*, as well as the African armyworm, *Spodoptera exempta* (Kaaya, unpublished data). Haemolymph is collected from the tsetse by amputation of the limbs of anaesthetized insects, followed by application of gentle pressure on the thorax with a pair of forceps. The haemolymph oozing from the stumps can then be aspirated into micropipettes. In lepidopteran larvae, anaesthesia is usually not necessary; incisions are made on the ventral abdominal cuticle. In addition to treatment with PTU, the haemolymph from adult insects must be centrifuged for 5 minutes at $10,000 × g$ to remove haemocytes and tissue debris.

Induction and Response to Infection

When live non-pathogenic or killed pathogenic bacteria are injected into an insect, they are rapidly removed from the circulation by haemocytes through phagocytosis and nodule formation (Christensen and Nappi, 1988; Dunn, 1986; Götz and Boman, 1985; Ratcliffe and Rowley, 1979; Ratcliffe *et al.*, 1985). In pupae of *S. cynthia*, most bacteria are cleared from the haemolymph immediately but some survive for up to one week in certain tissues. Granulocytes containing bacteria have been observed to aggregate with other haemocytes in the fat body and with the pericardial cells (Boman and Hultmark, 1987; Boman and Steiner, 1981). In the adult tsetse injected with *E. coli* D31, a low number of bacteria have also been observed to persist in the insect for 7-9 days (Kaaya, unpublished data). Thereafter, synthesis of immune RNA and specific immune proteins gave rise to antibacterial activity in the haemolymph. Although various non-pathogenic bacteria can induce immunity, *E. cloacae* was found to be the most effective vaccine (Faye *et al.*, 1975).

In *H. cecropia* and *A. pernyi*, injection of bacteria was followed by a lag period of 8-10 hours (Boman, 1982; Boman and Hultmark, 1987), after which levels of antibacterial activity in the haemolymph increased, peaking after 7-8 days. In the smaller *S. cynthia*, the peak was reached after 2-4 days (Boman and Hultmark, 1987; Faye *et al.*, 1975; Götz and Boman, 1985). In adult tsetse flies, the lag period was approximately 6 hours and peak antibacterial activity was reached after 30-48 hours (Kaaya, 1989a and b; Kaaya and Darji, 1988). In *Cecropia*, u.v. and heat-killed bacteria have been reported to stimulate very weak immune responses (Boman and Steiner, 1981), whereas in tsetse flies, heat-killed bacteria completely failed to stimulate production of cecropin and attacin-like antibacterial factors (Kaaya, 1989b; Kaaya and Darji, 1988). After the peak is reached, antibacterial activity gradually declines and disappears in about the same length of time required for reaching the maximum (Boman and Hultmark, 1987).

In silkworms, electrophoresis of the immune haemolymph on SDS-PAGE produced 9 bands of proteins, which were initially designated P1-P9 (Boman and Hultmark, 1981). Subsequently the proteins in bands P5 and P9 were named 'attacins' and 'cecropins' respectively, whereas P7 was identified as lysozyme (Boman and Hultmark, 1981, 1987; Hultmark *et al.*, 1983). The protein P9 was found to comprise two components, P9A and P9B, which were correspondingly labelled cecropins A and B (Boman and Steiner, 1981). Some of the 20-25 proteins produced in cecropians in response to bacterial infections were then purified and sequenced (Boman *et al.*, 1986; Boman and Hultmark, 1987; Kockum *et al.*, 1984).

Electrophoresis of the immune haemolymph in the tsetse fly on SDS-PAGE revealed only two new proteins with an MW of 17,000 and 70,000 daltons respectively. They were visible in stained gels of haemolymph collected 18 hours after immunization and peaked at 48 hr (Kaaya et al., 1986a and b). In larvae of *P. terranovae*, at least five proteins with MW of 3000-9000 daltons have been reported (Dimarcq *et al.*, 1986). In larvae of *S. peregrina*, one antibacterial protein (sarcotoxin) with an MW of 5,000 daltons has also been reported (Okada and Natori, 1983). In the tobacco hornworm, *M. sexta*, injection of bacteria led to the production of more than 25 different proteins (Hulbert *et al.*, 1985) and the antibacterial activity, as in cecropians, appeared to be due to the three usual families of antibacterial proteins: lysozyme, cecropins and attacins (Spies *et al.*, 1986a). Antibacterial factors have also been reported in several dipterans, e.g. in pupae (Bakula, 1970) and adult *Drosophila* (Boman *et al.*, 1972; Flyg *et al.*, 1987; Robertson and Postlethwait, 1986) and in a darkling beetle (Spies *et al.*, 1986b). Antibacterial factors of low molecular weight have been induced in locusts (Lambert and Hoffmann, 1985) and in the assassin bug (De Azambuja *et al.*, 1986).

Work with pupae of *S. cynthia* has shown that both actinomycin D and cycloheximide inhibit expression of immunity, indicating that *de novo* synthesis of RNA and proteins occurs. Studies with actinomycin D also showed that 5 hours is required for synthesis of immune RNA (Boman *et al.*, 1974; Boman and Steiner, 1981; Faye *et al.*, 1975). In the tsetse also, cycloheximide inhibited production of antibacterial activity (Kaaya *et al.*, 1987).

Site of Antibacterial Protein Synthesis

Lysozyme is present in the haemocytes of larvae of *Spodoptera eridania* and increases following injection of foreign materials (Anderson and Cook, 1979). In *Locusta migratoria*, lysozyme is localized in granular haemocytes (Zachary and Hoffmann, 1984). Synthesis of antibacterial proteins by cultured fat body cells from immunized pupae of *H. cecropia* has been reported (Faye and Wyatt, 1980). Upon electrophoresis, the proteins produced by these cells co-migrated with cecropins and lysozyme. The fat body of non-immunized larvae of *M. sexta* contained immunoreactive lysozyme that was released into a culture medium *in vitro* (Dunn and Drake, 1983). The lysozyme levels and release rates were notably higher in the fat body obtained from immunized larvae. The fat body of immunized *M. sexta* also released cecropin-like antibacterial activity *in vitro* (Dunn, 1986). These results and others (Boman and Hultmark, 1987) clearly indicate that the fat body must be the major site of synthesis for the immune proteins.

Techniques for Determination of Antibacterial Activity and Immune Proteins

STANDARD ASSAYS FOR ANTIBACTERIAL ACTIVITY

Killing assay: The techniques described here are slight modifications of those described earlier by other authors (Boman *et al.*, 1974; Engstrom *et al.*, 1984) for detection of antibacterial activity in the haemolymph of immunized pupae of silkworms. The standard buffer is 0.1 M sodium or potassium phosphate pH 6.4 with 2×10^{-3} M dithiothritol (DDT, 'Cleland's reagent'; Sigma Chemicals). The mixture contains 5 or 10 µl of 10-20 times diluted test haemolymph and 100 µl of standard buffer containing 5×10^6 test bacteria (*E. coli* D1 or *Bacillus subtilis* BS11) in a 1.5-ml conical polypropylene tube. Incubation is carried out at room temperature and, at short intervals, 5 µl samples are withdrawn from the reaction mixture, using disposable micropipettes, and emptied into 1 ml of ice-cold 0.9% NaCl to stop the killing reaction. After further dilution, the samples are spread on nutrient agar plates at 45 °C. For *E. coli* D31 the plates are supplemented with 80 µg of ampicillin or 100 µg of streptomycin per ml. After 30 minutes, a second layer of soft agar is spread on each plate; the plates are then incubated at 37°C overnight. This procedure allows an accurate counting of up to 1000 colonies per plate. The result of a 3-minute reaction may be used as a measure of the antibacterial activity (Boman *et al.*, 1974; Faye *et al.*, 1975). In small insects, such as members of *Drosophila*, those immunized with *E. cloacae* B12 and saline-injected controls are homogenized with 50 µl of 0.9% NaCl containing 100 µg of streptomycin per ml. Each sample is then inoculated with 4×10^4 viable *E. cloacae* B12, the mixture incubated at 30°C and, at different times, samples of 5 µl withdrawn, diluted and plated (Boman *et al.*, 1972; Flyg *et al.*, 1987).

Bacteriolytic assay: Log-phase *E coli*, strain D31, is centrifuged and suspended in ice-cold 0.1 M phosphate buffer, pH 6.4, to give a density of 30 units on a Klett-Summerson colorimeter (A_{570} 0.3-0.5). A small volume, e.g. 10 µl, of haemolymph sample (or buffer for control) is added to 1 ml of bacterial suspension in an ice-bath. The mixture is then incubated for 30 minutes at 37°C. The tubes are then transferred to an ice-bath and absorbance at 570 nm is measured within 1 hour after incubation. One unit of lytic activity is defined as the amount of factor giving a 50% reduction of absorbance at 570 nm compared to the control. The value for 50% lysis may be obtained by interpolation, or alternatively the number of units (U) is estimated by the following formula obtained empirically:

$$U = \sqrt{\frac{Ao - A}{A}}$$

where   A is the absorbance in the test sample, and Ao the absorbance in the control.

Using this expression the number of units can be estimated from a single reading between 30 and 70% lysis (Hultmark *et al.*, 1980).

Inhibition zone assay: A slight modification of the method of Hoffmann *et al.* (1981) is described. Plates are prepared by adding log-phase *E. coli* D31 to 7 ml of 1% nutrient agar at 45°C, containing 100 µg streptomycin/ml, to a final concentration of 10^5 bacteria/ml. After mixing, 6 ml of agar is placed in a 9 cm plastic disposable Petri dish and left for 30 minutes to solidify. Two-mm diameter wells are made and 2 µl of test haemolymph are placed in each well, followed by overnight incubation at 37°C. Once prepared, plates can be stored in a refrigerator at 4°C for up to 2 weeks. A standard curve can be prepared for each experiment by including a dilution series of a proper reference. Diameters of the growth inhibition zones (Fig. 1) are recorded for the test samples as well as the reference.

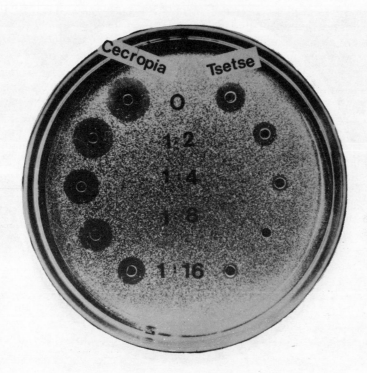

Fig. 1: Inhibition-zone assay demonstrating antibacterial activity in serially diluted cecropians and tsetse immune haemolymphs. Two µl samples were put in wells in thin agar seeded with *E. coli* D31. Notice bacteria-free zones around wells.

There is a linear relationship between the diameter of the bacteria-free zone and the logarithm of the concentration of the reference used.

Lysozyme Assay

For detection of lysozyme activity, two types of tests have been described (Hoffman *et al.*, 1981; Hultmark *et al.,* 1980; Zachary and Hoffmann, 1984).

Inhibition zone assay: This technique is very similar to the one described for antibacterial activity above. Six ml of 1% agar in 0.1 M phosphate buffer, pH 6.4, containing 1 mg/ml lyophilized *M. luteus* (Sigma chemicals) is placed in sterile disposable 9-cm Petri dishes. After solidification, 2-mm diameter wells are made and 2 µl test samples are placed in each well. Simultaneously, samples of serial dilutions of chicken egg-white lysozyme with known concentrations are placed in wells on the same plate, incubated at 30°C overnight and the diameters of the clear zones around the wells recorded. A standard curve using diameters of the clear zones around the wells containing the chicken egg-white lysozyme is prepared, from which lysozyme concentrations in the haemolymph samples can be estimated

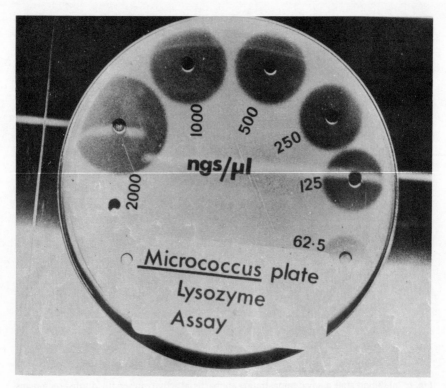

Fig. 2: Lysozyme assay showing clear zones around wells in agar seeded with dry lyophilized *M. luteus*. Two μl samples of serially diluted chicken egg-white lysozyme of known concentrations were placed in each well.

(see Fig. 2). As for the antibacterial activity, there is a linear relationship between the diameter of the clear zones and the logarithm of lysozyme concentration (Mohrig and Messner, 1968a).

Nephelometric test: Dried *M. luteus* cells (Sigma) are suspended in cold 0.1 M phosphate buffer, pH 6.4, to give a density of 0.2-0.3 on a Bausch and Lomb Spectronic 20, at 570 nm (Hultmark *et al.*, 1980). Ten μl of each sample to be tested is then added to 1 ml of bacterial suspension in phosphate buffer, in an ice-bath, after which the mixture is incubated for 30 min at 37°C. The reaction is stopped by placing the suspension in an ice-bath. The absorbance at 570 nm is measured and then compared to those of reference samples containing known amounts of standard chicken egg-white lysozyme.

Sodium-dodecyl-sulphate Polyacrylamide Gel Electrophoresis (SDS-PAGE).

Polyacrylamide slab gel electrophoresis is performed with an electrophoresis apparatus (Pharmica Fine Chemicals, Uppsala, Sweden) at a gel concentration of 7.5%. The electrophoresis buffer is 0.1 M sodium phosphate containing 0.1% sodium dodecyl sulphate. The pH of the electrode buffer should be 6.4 while that of the gel buffer is 7.2. Some lanes in the gel are loaded with 10-15 μl of test haemolymph and others with molecular weight markers. Electrophoresis is run at 150 v for approximately 3 hours, after which the gels are stained overnight using 0.1% Coomassie brilliant blue and then destained in 7% glacial acetic acid until the background is clear (Faye *et al.*, 1975; Kaaya *et al.*, 1986a; Kaaya *et al.*, 1986b). Electrophoretic mobilities of the molecular weight markers (e.g. phosphorylase, bovine serum albumin, ovalbumin) allow a fairly accurate estimation of the molecular weights of the haemolymph proteins (Fig. 3).

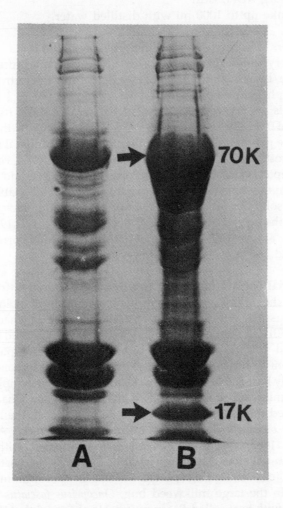

Fig. 3: SDS-PAGE of tsetse haemolymph from saline-(A) and live *E. coli* D31-(B) injected adult tsetse fly. Notice the increased immune proteins (arrows) in the immune haemolymph (B).

Analytical Polyacrylamide Gel Electrophoresis for Demonstration of Attacins and Cecropins

Reagents:
(1) Electrophoresis buffer (10 × concentrated):
    — B-alanine 31.2 g.
    — 8.10 ml concentrated acetic acid.
    — Make up to 1000 ml with distilled water. Dilute 10 × when using.
(2) Gel buffer (2.5 M potassium acetate pH 3.9).
    — 143 ml concentrated acetic acid.
    — 800 ml distilled H$_2$O.
    — 38.9 g KOH, -stir.
    — Make up to 1000 ml with distilled water.

Procedure: Electrophoresis of immune haemolymph is carried out in 15% polyacrylamide gels at pH 4 using a discontinuous buffer system and acrylamide/bisacrylamide ratio of 60 : 0 : 8. Prior to electrophoresis, the haemolymph samples are acidified by adding 0.5 vol. of 1 M acetic acid and the gels are run at 200 v until the tracker dye starts running off the gel (2.5-3 hΩ) (Hultmark *et al.*, 1982). To localize bands with antibacterial activity, the gels are incubated for 1 hour in a rich bacterial medium (e.g. nutrient broth) in 0.2 M sodium phospate, buffer pH 7.4, containing 100 µg/ml streptomycin. The gels (8 × 8 cm) are then overlaid with 5 ml of agar 0.6% in the same medium containing about 2×10^5 viable *E. coli* D31. Soon after solidification another layer of agar without bacteria is poured on top and the gel is incubated at 37°C overnight (Flyg *et al.*, 1987; Hultmark *et al.*, 1980; Kaaya *et al.*, 1987). Spots without bacteria corresponding to attacins and cecropins can be seen on the gels (Fig. 4).

Antibacterial Factors in Insects Induced by Bacterial Infections

Among the earliest pioneers in the field of insect immunology are Briggs (1958), Gingsrich (1964), Stephens and Marshall (1962), Chadwick (1967), Bakula (1970) and Boman *et al.* (1972). Briggs (1958) observed that injection of live and attenuated bacteria into larvae of lepidopterans led to development of resistance paralleled by an increase in concentration of an extremely heat-stable antibacterial factor in the haemolymph a few hours after immunization. Stephens and Marshall (1962) and Chadwick (1967) also reported the presence of an antibacterial factor in the haemolymph of *Galleria mellonella* larvae immunized with bacteria. According to Chadwick (1967), larvae of *G. mellonella* vaccinated with heat-killed *P. aeruginosa* developed immunity within 24 hours that lasted for about 70 hours. Gingsrich (1964) reported development of immunity to live *P. aeruginosa* in the large milkweed bug, *Oncopeltus fasciatus* (Hemiptera) vaccinated with heat-killed *P. aeruginosa*. He observed that the immunity

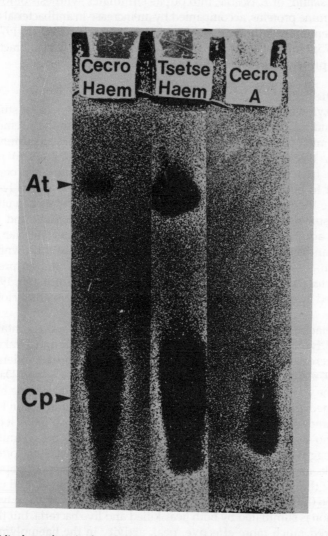

Fig. 4: Acidic electrophoresis of cecropian and tsetse immune haemolymphs as well as purified cecropin A from cecropians, followed by antibacterial assay (soft agar seeded with *E. coli* D31) on the gel. Notice positions of attacins (At) and cecropins (Cp) with no bacterial growth.

was associated with the presence of lytic substances in the haemolymph and that both the immunity and the lytic substances appeared 4 hours after vaccination, peaked at 2 days and disappeared within 5 days. Bakula (1970) and Boman *et al.* (1972) reported induction of antibacterial activity in pupal and adult stages of *Drosophila melanogaster* respectively and, to my knowledge, are the first authors to report immunity in a dipteran.

More recently, comprehensive studies have been conducted in diapausing pupae of silkmoths by Boman and his associates in Sweden.

Their studies showed that injection of live non-pathogenic bacteria, e.g.
E. coli, B. subtilis or *E. cloacae,* into pupae stimulates synthesis of RNA and
20-25 immune proteins, accompanied by an increase in antibacterial activity
in the haemolymph (Boman and Hultmark, 1981; Boman *et al.*, 1974; Faye
et al., 1975; Hultmark *et al.*, 1980; Kirschbaum, 1985). The characteristics
of these proteins are briefly described below:

LYSOZYMES
    Lysozyme was the first humoral antibacterial factor to be studied in
insects and was considered the main immune factor in the haemolymph
(Mohrig and Messner, 1968a) as well as in the gut of insects (Mohrig and
Messner, 1968b). Later, Powning and Davidson (1973, 1976) purified
lysozyme from haemolymph of *G. mellonella,* the first immune protein from
an insect to be isolated in a pure form. It is now known that lysozyme
cannot be the main immune factor in insects because they can eliminate
many lysozyme-resistant bacteria (Boman *et al.*, 1986; Götz and Boman,
1985). In addition to its presence in the gut (Mohrig and Messner, 1968b;
Ribeiro and Pereira, 1984) and haemolymph of insects (Powning and
Davidson, 1973, 1976), lysozyme has also been reported as present in
haemocytes (Anderson and Cook, 1979; Zachary and Hoffmann, 1984) and
also stored in granules of the granular haemocytes (Zachary and Hoffmann,
1984).
    The haemolymph of non-immunized insects normally contains low
levels of lysozyme (Anderson and Cook, 1979; Dunn and Drake, 1983)
which increase manifold following infection with bacteria (Anderson and
Cook, 1979; Chadwick, 1970; Kaaya *et al.*, 1987; Powning and Davidson,
1973). The complete amino acid sequence of cecropia lysozyme has been
determined and a cDNA clone containing lysozyme information has been
isolated and sequenced (Boman *et al.*, 1986; Engstrom *et al.*, 1985). It contains
120 amino acids, has MW of 13,800 daltons and shows great similarity to
chicken egg-white lysozyme. Purification of cecropian lysozyme often gives
two peaks and structural studies have shown that cecropian populations
comprise three variants: (i) Arg-15, Ser-66 (ii) Leu-15, Ser-66 and (iii) Leu-
15, Thr-66 (Boman *et al.*, 1986). In the adult tsetse *G. m. morsitans*, lysozyme
production is stimulated by both heat-killed and live bacteria, but the latter
has proved much more effective; peak activity in the haemolymph was
reached 1-2 hr post-immunization (Kaaya, 1989a; Kaaya and Darji, 1988;
Kaaya *et al.*, 1987).
    Unlike cecropins and attacins that are inhibited by cycloheximide, an
inhibitor of eukaryotic protein synthesis (see relevant sections below),
injection of cycloheximide into adult tsetse led to a rapid increase in
lysozyme levels in the haemolymph (Kaaya, 1989b; Kaaya *et al.*, 1987). It
is therefore likely that cycloheximide lyses the granular haemocytes, thus
releasing the lysozyme stored in these haemocytes (Zachary and Hoffmann,
1984).

CECROPINS

The cecropins are a family of inducible, strongly basic antibacterial proteins with MW of approximately 4000 daltons. They are active against both gram-positive and gram-negative bacteria. They can be demonstrated in immune haemolymph by acidic electrophoresis at pH 4 in combination with an antibacterial assay. The cecropins, and also attacins (discussed later), are visualized as spots without bacterial growth (Boman and Steiner, 1981; Hultmark et al., 1980; Kaaya et al., 1987). Using this technique, cecropin-like molecules have also been demonstrated in various lepidopterans (Hoffmann et al., 1981), Drosophila (Flyg et al., 1987; Robertson and Postlethwait, 1986), tsetse flies (Kaaya et al., 1987) and in darkling beetles (Spies, et al., 1986b).

In cecropians, the cecropins are a family of six antibacterial proteins, designated A-F with A, B, D the major ones; in A. pernyi, cecropin D is the major one and A and B are present at lower concentrations (Boman and Hultmark, 1987; Götz and Boman, 1985; Qu et al., 1982). Cecropins A, B and D from H. cecropia, B and D from A. pernyi and other cecropins similar to the A and B in Bombyx mori have been purified and sequenced (Boman et al., 1986; Boman and Hultmark, 1987). A cecropin-like antibacterial factor (sarcotoxin 1) from S. peregrina has also been purified and sequenced (Okada and Natori, 1983). These sequence studies have shown clearly that sarcotoxin 1 is a cecropin with about 40% homology to cecropin A from cecropians (Boman et al., 1986).

The amino acid sequences of the major cecropins have been studied (Boman et al., 1986; Boman and Steiner, 1981). The N-terminal parts are strongly basic, whereas the C-terminal regions are neutral and contain long hydrophobic stretches. In all cases, the cecropins have an amidated C-terminal residue (Boman and Hultmark, 1987). In cecropians, the B and D forms show 65% and 62% homology respectively with cecropin A. About half the amino acid substitutions are strictly conservative and hence the various cecropin molecules are very similar. The high degree of homology among cecropins A, B, D suggests that they have arisen through gene duplication of a common ancestral gene (Boman and Hultmark, 1987; Götz and Boman, 1985).

Melittin, the bee venom toxin, is active against E. coli, Bacillus megaterium and B. subtilis (Boman, 1982). The general design of melittin is similar to that of the cecropins but the polarity is reversed; thus in melittin the C-terminal end is basic and the N-terminal end is hydrophobic. Moreover, both proteins have a proline in the middle, a single tryptophan in front of a basic sequence stretch and amidated C-terminals formed from glycine residues (Boman and Hultmark, 1987). Cecropin A lyses E. coli but not Chang liver cells, even at a concentration 300 times higher than needed for killing bacteria. Melittin lyses both bacteria and Chang liver cells (Boman, 1982; Götz and Boman, 1985; Steiner et al., 1981). Cecropins have no effect on other mammalian cell lines or on yeast. They appear to be specific for prokaryotic cells, whereas melittin acts on both prokaryotic and eukaryotic cells (Boman and Hultmark, 1987; Götz and Boman, 1985). It seems that the main targets of the cecropins are bacterial membranes, whereas lysozyme acts on the murein skeletons, implying that both antibacterial proteins are needed for complete and effective destruction of bacteria (Götz and Boman, 1985).

ATTACINS

Attacins were first isolated as protein P5 from immune haemolymph of *H. cecropia* (Boman and Hultmark, 1981; Kockum *et al.*, 1984) and were later found to be a family of six (A-F) closely related antibacterial proteins that could be fractionated according to isoelectric point (Boman, 1986; Boman and Hultmark, 1987; Hultmark *et al.*, 1983). They have isoelectric points ranging from 5.7 to 8.3 and an MW of 20,000-23,000 daltons (Hultmark *et al.*, 1983). Five different attacin-like peptides have also been detected in *M. sexta* (Hurlbert *et al.*, 1985), whereas attacin-like antibacterial factors have been reported in tsetse flies (Kaaya *et al.*, 1987) and in *S. peregrina* (Ando *et al.*, 1987). Four attacins, A-D, are basic while E and F are acidic (Hultmark *et al.*, 1983; Kockum *et al.*, 1984). The N-terminal sequences for three of the basic forms are identical, whereas the two acidic forms have a slightly different sequence (Boman and Hultmark, 1987). The two genes for attacins, one for basic and another for acidic (Kockum *et al.*, 1984), may have arisen, like those of cecropins, through gene duplication. Attacin E and F have 188 and 184 amino acid residues respectively (Boman and Hultmark, 1987; Engstrom *et al.*, 1984).

The antibacterial spectrum of attacins appears to be rather narrow, with high activity only against *E. coli* and two other gram-negative bacteria, *Acinetobacter calcoaceticus* and *Pseudomonas maltophilia*, the latter two originating from the gut of *A. pernyi* larvae (Boman, 1986; Hultmark *et al.*, 1983). A study of the mechanisms of action on *E. coli* revealed that attacins act on the outer membrane and that they facilitate the action of cecropins and lysozyme, thus enabling the three immune proteins to work in consonance (Boman and Hultmark, 1987).

Other Anti-bacterial Factors

It has been demonstrated that larvae of the dipteran insect *P. terranovae* synthesize several peptides with potent antibacterial activity in response to bacterial challenge or injury (Keppi *et al.*, 1986). Three of these peptides have already been isolated and their amino acid sequences determined (Dimarcq *et al.*, 1988; Keppi *et al.*, 1989). They are basic molecules, heat-stable, with molecular weight of 8,600 daltons and have been termed diptericins (Dimarcq *et al.*, 1986; Dimarcq *et al.*, 1988). More recently, another family of inducible antibacterial peptides, termed defensins, has been reported in immunized larvae of *P. terranovae* (Dimarcq *et al.*, 1990).

Whereas the diptericins are active against gram-negative bacteria, the defensins act against gram-positive bacteria and show significant homology to mammalian microbicidal peptides present in polymorphonuclear leukocytes and macrophages (Dimarcq *et al.*, 1990). A defensin-related antibacterial peptide, referred to as sapecin, has also been reported in *S. peregrina* (Matsuyama and Natori, 1988). Two cell types, fat body cell and thrombocytoid, have been shown to produce both diptericins and defensins in *P. terranovae* (Dimarcq *et al.*, 1990).

The Injury Reaction

In their studies on immunity in silkworm pupae, Boman *et al.* (1974) observed that a sham-injection with a sterile salt solution induced a low level of antibacterial activity, which was interpreted as an injury effect caused by the injection needle (Boman and Steiner, 1981). Even an injury as small as that caused by a fine needle suffices for significant stimulation of protein synthesis (Faye *et al.*, 1975, Götz and Boman, 1985). Proteins P1, P4, P5, P7 and P8 were found to be present in reduced amounts or were even absent in the injury response caused by the injection of sterile saline (Faye *et al.*, 1975). The main component in the injury response, P3, was only a minor fraction in the response obtained with *E. cloacae*. Significantly, the immune response always contains potent antibacterial activity, whereas the injury response induces either no or only weak antibacterial activity (Götz and Boman, 1985). In insects, wounds are likely entry points for infections and it is therefore not surprising that injuries activate the humoral immune system (Boman and Hultmark, 1987).

Passive and Active Resistance to Immunity

Two main types of avoidance mechanisms have been found to counteract insect immunity, namely, 'passive' and 'active' resistance (Boman and Hultmark, 1987; Boman and Steiner, 1981). Passive resistance occurs if the envelope of a bacterium can prevent the insect's antibacterial activity from killing the bacterium. For instance, a barrier in the cell wall makes *B. thuringiensis* totally insensitive to both cecropins and attacins. Passive resistance has also been reported in the insect-pathogenic bacterium, *Serratia marcescens* (Boman and Steiner, 1981; Götz and Boman, 1985).

Active resistance, on the other hand, is due to destruction of immunity by an active process, e.g. by proteolytic digestion of cecropins and attacins by 'immune inhibitors'. The first such inhibitor was isolated from *B. thuringiensis* and is known as In-A (Dalhammar and Steiner, 1984). It is a proteolytic enzyme consisting of a single polypeptide chain with MW of 78,000 daltons (Boman, 1982). It degrades both attacins and cecropins of *H. cecropia* (Dalhammar and Steiner, 1984) and the tsetse fly (Kaaya *et al.*, 1987) (see Fig. 5).

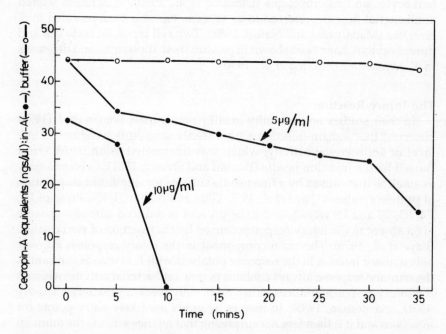

Fig. 5: Effect of Inhibitor A (In-A) from *Bacillus thuringiensis* on antibacterial factors in tsetse immune haemolymph. Two dilutions of purified In-A (5 and 10 μg/ml (•) and control (o) were preincubated with immune haemolymph and buffer respectively for different times. Notice the inactivation of the antibacterial activity in the In-A-containing immune haemolymph.

The nematode, *Neoaplectana carpocapsae*, which lives in a symbiotic relationship with a gram-negative bacterium (*Xenorhabdus nematophilus*), also produces an immune inhibitor. This A-type inhibitor destroys both cecropins and attacins, thus protecting the bacterium from destruction by insect immune factors; the nematode depends on this bacterium for killing the insect host (Boman, 1982; Götz and Boman, 1985). *S. marcescens* has been reported to produce A-type inhibitors (two proteolytic enzymes) that inhibit insect immunity (Boman, 1982).

Specificity
The specificity of insect immunity appears to be very broad and mammalian immunologists may even claim that it is lacking altogether (Boman and Steiner, 1981). For instance, immunization of *H. cecropia* pupae with *E. cloacae* produced an antibacterial activity directed against a variety of both gram-negative and gram-positive bacteria (Boman and Steiner, 1981; Kaaya, 1989a). However, it is not correct to say that the insect immune system lacks specificity altogether because insects are able to distinguish

between live and killed bacteria (Boman and Steiner, 1981; Kaaya, 1989 a and b; Kaaya and Darji, 1988) and also between 'self' and 'non-self' materials.

Discussion and Conclusions

As mentioned above, insects possess at least three families of antibacterial proteins: lysozyme, cecropins and attacins (Boman and Hultmark, 1987; Hultmark *et al.*, 1983; Kirschbaum, 1985). Both the cecropin and attacin families comprise six subcomponents, designated A-F (Boman, 1986; Hultmark et al., 1983). Although each of these subcomponents may differ in its antibacterial spectra, no bacterium has been found on which only a single subcomponent acts. Furthermore, there is no evidence of variable synthesis of specific proteins in response to different bacterial challenges, whether gram-positive or gram-negative. Why then should an insect such as *H. cecropia* have so many factors performing the same function? In attempting to answer this question, Boman (1986) suggested three possibilities: (i) the different factors may simply represent proteins 'in the middle' of an evolution towards separate functions, (ii) they may have separate target organisms which have not yet been identified and (iii) each subcomponent might have a separate target on most of the organisms on which these factors act. The latter is considered the most likely possibility. If such is the case, this would be of great survival value to the insect because it would be almost impossible for a susceptible bacterium to produce mutants resistant to the humoral immunity of the animal.

Lysozyme, the third family of insect antibacterial proteins kills only a few gram-positive bacteria. To date, no bacterium has been found that is lysozyme-sensitive and cecropin-resistant. Thus, it has been suggested that the main function of lysozyme may not be to kill bacteria, but to remove the murein succulus left behind after the action of cecropins and attacins (Boman, 1986; Boman and Hultmark, 1987).

Studies conducted in larvae of *M. sexta* have shown that following inoculation of non-pathogenic bacteria, e.g. *E. coli*, more than 90% of the bacteria were killed within 2 hours post-inoculation by cellular reactions, i.e., phagocytosis and nodule formation. The remaining bacteria were killed more slowly over a period of 30-40 hours (Dunn and Drake, 1983). Antibacterial proteins appeared in the haemolymph at 8 hours post-inoculation when most of the bacteria had already been killed. When a pathogen, e.g. *P. aeruginosa*, was injected, the same initial rapid killing by the haemocytes occurred, but at 8-10 hours the bacteria increased rapidly and killed the insects. From these studies, it was suggested that the role of haemocytes might be to detect the invader and to reduce its numbers rapidly. The synthesis of antibacterial proteins provides a back-up defensive when the initial inoculum is large, for 'mopping up' the remaining enemy.

This hypothesis is consistent with observations on the dose-response relationship between the number of bacteria inoculated and the levels of antibacterial protein synthesis (Kaaya *et al.*, 1987). In this case, the observed minimal threshold for induction would correspond to levels of infection that haemocytes alone can eliminate (Dunn, 1986).

The broad-spectrum nature of the antibacterial proteins produced in response to any bacterial infection and the short duration of the response, definitely distinguish it from the phenomenon of immunity in vertebrates. Thus Dunn (1986) has speculated that, in addition to aiding in complete removal of the invader, the synthesis of the broad-spectrum antibacterial proteins may serve to protect the disarmed insect while its primary defensive arsenal, the haemocytes, is being replenished, because the initial haemocyte response depletes the haemocytes dramatically (Ratcliffe and Rowley, 1979).

The immune system of insects is stimulated by non-pathogenic bacteria as well as by a few pathogenic ones. This led Boman and Hultmark (1987) to propose that the main function of the insect immune system is the control of the natural flora that insects come across in nature. This is a very important feature in such insects as houseflies and cockroaches which live in decomposing organic matter. However, as a result of long association between bacteria and insects, some pathogenic bacteria have evolved that circumvent the immune system.

Despite differences of opinion amongst various investigators, it is obvious that insects have the capacity to recognize and destroy foreign invaders by means of phagocytosis, nodule formation, encapsulation, melanization and the antibacterial proteins. If, therefore, one uses the commonly accepted definition of immunity, i.e., the ability of an organism to resist diseases, and discounts the requirement for a high degree of specificity and immunological memory characteristic of vertebrate immunity, then insects do indeed possess a true and effective immune system (Christensen and Nappi, 1988).

REFERENCES

Anderson, R.S. and M.L. Cook. 1979. Induction of lysozymelike activity in the hemolymph and hemocytes of an insect, *Spodoptera eridania*. *J. Invertebr. Pathol.* 33: 197-203.

Ando, K., M. Okada and S. Natori. 1987. Purification of sarcotoxin II, antibacterial proteins from *Sarcophaga peregrina* (flesh fly) larvae. *Biochem.* 26: 226-230.

Bakula, M. 1970. Antibacterial compounds in the cell-free haemolymph of *Drosophila melanogaster*. *J. Insect Physiol.* 16: 185-197.

Boman, H.G. 1982. Humoral immunity in insects and the counter defence of some pathogens. *Fortschr. Zool.* 27: 211-222.

Boman, H.G. 1986. Antibacterial immune proteins in insects. *Symp. Zool. Soc. Lond.* 56: 45-58.

Boman, H.G. and D. Hultmark. 1981. Cell-free immunity in insects. *Trends Biochem. Sci.* 6: 306-309.

Boman, H.G. and H. Steiner. 1981. Humoral immunity in Cecropia pupae. In: *Current Topics in Microbiology and Immunology*, 94/95:75-91. (W. Henle *et al.*, eds.). Springer-Verlag, Berlin-New York.

Boman, H.G. and D. Hultmark. 1987. Cell-free immunity in insects. *Ann. Rev. Microbiol.* 41: 103-126.

Boman, H.G., I. Nilsson and B. Rasmuson. 1972. Inducible antibacterial defense system in *Drosophila. Nature* 237: 232-235.

Boman, H.G., I. Nilsson-Faye and T. Rasmuson. 1974. Why is insect immunity interesting? In: *Energy, Biosynthesis and Regulation in Molecular Biology*, pp. 103-114 (D. Richter, ed.). Walter de Gruyter, Berlin.

Boman, H.G., I. Nilsson-Faye, K. Paul and T. Rasmuson. 1974. Insect immunity. I. Characteristics of an inducible cell-free antibacterial reaction in hemolymph of *Samia cynthia* pupae. *Inf. Immun.* 10: 136-145.

Boman, H.G., I. Faye, A. Pye and T. Rasmuson. 1978. The inducible immune system of giant silk moths. In: *Comp. Pathobiol.*, pp. 145-163 (L.E. Bulla and T.C. Cheng, eds.). Plenum Publishing Corporation, New York.

Boman, H.G., I. Faye, P.V. Hofsten, K. Kockum, J.Y. Lee, K.G. Xanthopoulos, H. Bennich, A. Engstrom, B.R. Merrifield and D. Andrech. 1986. Antibacterial immune proteins in insects—A review of current perspectives. In: *Immunity in Invertebrates*, pp. 63-73 (M. Brehelin, ed.). Springer-Verlag, Berlin-Heidelberg.

Briggs, J.D. 1958. Humoral immunity in lepidopterous larvae. *J. Exp. Zool.* 138: 155-158.

Chadwick, J.S. 1967. Serological response of insects. *Fed. Proc.* 26: 1675-1679.

Chadwick, J.S. 1970. Relation of lysozyme concentration to acquired immunity against *Pseudomanas aeruginosa* in *Galleria mellonella. J. Invertebr. Pathol.* 15: 455-456.

Christensen, B.M. and A.J. Nappi. 1988. Immune responses of arthropods. *ASI Atlas Sci.* 0894-3761: 15-19.

Dalhammar, G. and H. Steiner. 1984. Characterization of inhibitor A, a protease from *Bacillus thuringiensis* which degrades attacins and cecropins, two classes of antibacterial proteins in insects. *Eur. J. Biochem.* 139: 247-252.

De Azambuja, P., C.C. Freitas and E.S. Garcia. 1986. Evidence and partial characterization of an inducible antibacterial factor in the haemolymph of *Rhodnius prolixus. J. Insect Physiol.* 32: 807-812.

Dimarcq, J.L., E. Keppi, J. Lambert, D. Zachary and D. Hoffmann. 1986. Diptericin A, a novel antibacterial peptide induced by immunization or injury in larvae of the dipteran insect *Phormia terranovae. Dev. Comp. Immunol.* 10: 626.

Dimarcq, J.L., D. Zachary, J.A. Hoffmann, D. Hoffmann and J.M. Reichhart. 1990. Insect immunity: Expression of the two major inducible antibacterial peptides, defensin and diptericin in *Phormia terranovae. EMBO J.* 9: 2507-2515.

Dimarcq, J.L., E. Keppi, B. Dunbar, J. Lambert, J.M. Reichhart, D. Hoffmann, S.M. Rankine, J.E. Fothergill and J.A. Hoffmann. 1988. Purification and characterization of a family of novel inducible antibacterial proteins from immunized larvae of the dipteran *Phormia terranovae* and complete amino acid sequence of the predominant member, Diptericin A. *Eur. J. Biochem.* 171: 17-22.

Dunn, P.E. 1986. Biochemical aspects of insect immunology. *Ann. Rev. Entomol.* 31: 321-329.

Dunn, P.E. and D.R. Drake. 1983. Fate of bacteria injected into naive and immunized larvae of the tobacco hornworm, *Manduca sexta. J. Invertebr. Pathol.* 41: 77-85.

Engstrom, P., A. Carlsson, A. Engstrom, Z.J. Tao and H. Bennich. 1984. The antibacterial effect of attacins from the silk moth *Hyalophora cecropia* is directed against the outer membrane of *Escherichia coli. EMBO J.* 3: 3347-3351.

Engstrom, A., K.G. Xanthopoulos, H.G. Boman and H. Bennich. 1985. Amino acid and cDNA sequences of lysozyme from *Hyalophora cecropia. EMBO J.* 4: 2119-2122.

Faye, I. and Wyatt, G.R. 1980. The synthesis of antibacterial proteins in isolated fat body from cecropia silkmoth pupae. *Experientia* 36: 1325-1326.

Faye, I., A. Pye, T. Rasmuson, H.G. Boman and I.A. Boman. 1975. Insect immunity: II. Simultaneous induction of antibacterial activity and selective synthesis of some haemolymph proteins in diapausing pupae of *Hyalophora cecropia* and *Samia cynthia*. *Infect. Immun.* 12: 1426-1438.

Flyg, C., G. Dalhammar, B. Rasmuson and H.G. Boman. 1987. Insect immunity: Inducible antibacterial activity in *Drosophila*. *Insect Biochem.* 17: 153-160.

Gingsrich, R.E. 1964. Acquired humoral immune response of the large milkweed bug, *Oncopeltus fasciatus* (Dallas), to injected materials. *J. Insect Physiol.* 10: 179-194.

Götz, P. and H.G. Boman. 1985. Insect immunity. In: *Comprehensive Insect Physiology, Biochemistry and Pharmacology*, pp. 453-485. (G.A. Kerkut and L.I. Gilbert, eds.). Pergamon, Oxford-New York.

Hoffmann, D., D. Hultmark and H.G. Boman. 1981. Insect immunity: *Galleria mellonella* and other lepidoptera have Cecropia-P9-like factors active against gram-negative bacteria. *Insect Biochem.* 11: 537-548.

Hultmark, D., H. Steiner, T. Rasmuson and H.G. Boman. 1980. Insect immunity: Purification and properties of three inducible bactericidal proteins from haemolymph of immunized pupae of *Hyalophora cecropia*. *Eur. J. Biochem.* 106: 7-16.

Hultmark, D., A. Engstrom, H. Bennich, R. Kapur and H.G. Boman. 1982. Insect immunity: Isolation and structure of cecropin D and four minor antibacterial components from cecropia pupae. *Eur. J. Biochem.* 127: 207-217.

Hultmark, D., A. Engstrom, K. Andersson, H. Steiner, H. Bennich and H.G. Boman. 1983. Insect immunity: Attacins, a family of antibacterial proteins from *Hyalophora cecropia*. *EMBO. J.* 2: 571-576.

Hurlbert, R.E., J.E. Karlinsey and K.D. Spence. 1985. Differential synthesis of bacteria-induced proteins of *Manduca sexta* larvae and pupae. *J. Insect Physiol.* 31: 205-215.

Kaaya, G.P. 1989a. A review of the progress made in recent years on research and understanding of immunity in insect vectors of human and animal diseases. *Insect Sci. Applic.* 10: 751-769.

Kaaya, G.P. 1989b. Assessment of antibiotic potentials of insect antibacterial factors. *Insect Sci. Applic.* 10: 341-346.

Kaaya, G.P. and N. Darji. 1988. The humoral defense system in tsetse: Differences in response due to age, sex and antigen types. *Dev. Comp. Immunol.* 12: 255-268.

Kaaya, G.P., C. Flyg and H.G. Boman. 1987. Insect immunity: Induction of cecropin and attacin-like antibacterial factors in the haemolymph of *Glossina morsitans morsitans*. *Insect Biochem.* 17: 309-315.

Kaaya, G.P., L.H. Otieno, N. Darji and P. Alemu. 1986a. Defence reactions of *Glossina morsitans morsitans* against different species of bacteria and *Trypanosoma brucei brucei*. *Acta Trop.* 43: 31-42.

Kaaya, G.P., N.A. Ratcliffe and P. Alemu. 1986b. Cellular and humoral defenses of *Glossina* (Diptera: Glossinidae): Reactions against bacteria, trypanosomes, and experimental implants. *J. Med. Entomol.* 23: 30-43.

Keppi, E., D. Zachary, M. Robertson, D. Hoffmann and J.A. Hoffmann. 1986. Induced antibacterial proteins in the haemolymph of *Phormia terranovae* (Diptera): Purification and possible origin of one protein. *Insect Biochem.* 16: 395-402.

Keppi, E., A.P. Pugsley, J. Lambert, C. Wicker, J.L. Dimarcq, J.A. Hoffmann and D. Hoffmann. 1989. Mode of action of diptericin A, a bactericidal peptide induced in the hemolymph of *Phormia terranovae* larvae. *Ach. Ins. Biochem. Physiol.* 10: 229-239.

Kirschbaum, J.B. 1985. Potential implication of genetic engineering and other biotechnologies to insect control. *Ann. Rev. Entomol.* 30: 51-70.

Kockum, K., I. Faye, P. Von Hofsten, J.Y. Lee, K.G. Xanthopoulos and H.G. Boman. 1984. Insect immunity: Isolation and sequence of two cDNA clones corresponding to acidic and basic attacins from *Hyalophora cecropia*. *EMBO J.* 3: 2071-2075.

Komano, H., D. Mizuno and S. Natori. 1980. Purification of lectin induced in the hemolymph of *Sarcophaga peregrina* larvae on injury. *J. Biol. Chem.* 255: 2919-2924.

Lambert, J. and D. Hoffmann. 1985. Miseen evidence d'un facteur actif contre des bacteries gram negatives dans le sang de *Locusta migratoria*. *C.R.S. Acad. Sci.* (Paris) Ser. III, 300: 425-430.

Matsuyama, K. and S. Natori. 1988. Molecular cloning of cDNA for sapecin and unique expression of the sapecin gene during the development of *Sarcophaga peregrina. J. Biol. Chem.* 236. 17117-17121.

Mohrig, W. and B. Messner. 1968a. Immunreaktionen bei Insekten: I-Lysozym als grundlegender antibacterieller Factor im humoralem. *Abwehrmechanismus der Insecten. Biol. Zbl.* 87: 439-470.

Mohrig, W. and B. Messner. 1968b. Immunreaktionen bei Insekten: II. Lysozym als antimikrobielles Agens im Darmtrakt von Insekten. *Biol. Zbl.* 87: 705-718.

Okada, M. and S. Natori. 1983. Purification and characterization of an antibacterial protein from haemolymph of *Sarcophaga peregrina* (fleshfly) larvae. *J. Biochem.* 211: 727-734.

Powning, R.F. and W.J. Davidson. 1973. Studies on insect bacteriolytic enzymes. I. Lysozyme in haemolymph of *Galleria mellonella* and *Bombyx mori. Comp. Biochem. Physiol.* 45B: 669-686.

Powning, R.F. and W.J. Davidson. 1976. Studies on insect bacteriolytic enzymes II. Some physical and enzymatic properties of lysozyme from haemolymph of *Galleria mellonella. Comp. Biochem. Physiol.* 55B: 221-228.

Qu, X.-M., H. Steiner, A. Engstrom, H. Bennich and H.G. Boman. 1982. Insect immunity: Isolation and structure of cecropins B and D from pupae of the Chinese oak silk moth, *Antheraea pernyi. Eur. J. Biochem.* 127: 219-224.

Rasmuson, T. and H.G. Boman. 1979. Insect immunity: V. Purification and some properties of immune protein P4 from haemolymph of *Hyalophora cecropia* pupae. *Insect Biochem.* 9: 259-264.

Ratcliffe, N.A. and A.F. Rowley. 1979. Role of hemocytes in defence against biological agents. In: *Insect Hemocytes, Development, Forms, Functions and Techniques*, pp. 331-414 (A.P. Gupta, ed.). Cambridge University Press, Cambridge.

Ratcliffe, N.A., A.F. Rowley, S.W. Fitzgerald and C.P. Rhodes. 1985. Invertebrate immunity: Basic concepts and recent advances. *Int. Rev. Cytol.* 97: 183-350.

Ribeiro, J.M.C. and M.E.A. Pereira. 1984. Midgut glycosidases of *Rhodnius prolixus. Insect Biochem.* 14: 103-108.

Robertson, M. and J.H. Postlethwait. 1986. The humoral antibacterial response of *Drosophila* adults. *Dev. Comp. Immunol.* 10: 167-179.

Salt, G. 1970. The Cellular Defence Reactions of Insects. Cambridge Monograph in Experimental Biology, No. 16. Cambridge University Press, London.

Spies, A.G., J.E. Karlinsey and K.D. Spence. 1986a. Antibacterial hemolymph proteins of *Manduca sexta. Comp. Biochem. Physiol.* 83B: 125-133.

Spies, A.G., J.E. Karlinsey and K. Spence. 1986b. The immune proteins of the darkling beetle, *Eleodes* (Coleoptera: Tenebrionidae). *J. Invertebr. Pathol.* 47: 234-235.

Steiner, H., D. Hultmark, A. Engstrom, H. Bennich and H.G. Boman. 1981. Sequence and specificity of two antibacterial proteins involved in insect immunity. *Nature* 292: 246-248.

Stephens, J.M. and J.H. Marshall. 1962. Some properties of an immune factor isolated from the blood of actively immunized wax moth larvae. *Can. J. Microbiol.* 8: 719-725.

Zachary, D. and D. Hoffmann. 1984. Lysozyme is stored in the granules of certain haemocyte types in *Locusta. J. Insect Physiol.* 30: 405-411.

CHAPTER 7

Methods for Genetic Investigation of Cellular Immune Reaction in Insects, with the Parasitic Wasp-*Drosophila* System as a Model

Y. Carton and A. Nappi

Introduction

Until recently, investigations of the genetic aspects of the cellular immune reactions of insects and arthropods were scant (Götz and Boman, 1985; Stoffolano, 1986). One reason behind this paucity is that for most host-parasite interactions little or no information is available on the genetic systems of the interacting species. Also, it is extremely difficult in some host-parasite relationships to follow various genetic crosses under controlled laboratory conditions over several generations. Another reason for the lack of interest in studying the genetic aspects of insect immunity has been the notion that polygenic mechanisms were involved in insect host responses and that these parameters were suitable for genetic analyses. However, recent studies of the susceptibility of various insects to parasitic infections have shown that heritable variations exist in immune reactivity among natural populations of the same host species, especially in the *Drosophila*-parasitoid system (Bouletreau and Fouillet, 1982; Carton and Bouletreau, 1985; Carton, 1988a; Carton *et al.*, 1989), and that such variation can be studied and comprehensively analyzed for determining the degree of genetic influence. Knowledge of the genetic mechanisms influencing cellular immunity and parasite virulence (i.e., the capacity of the parasite to counteract or evade the immune reaction of the host) is fundamental to developing a comprehensive understanding of not only the physiological and biochemical aspects of these interacting responses (Nappi and Carton, 1986), but 'also of the co-evolutionary processes which operate at the populational level (Carton, 1988b; Carton and Nappi, 1991) and which impact significantly on biological control efforts.

This report provides an overview of the genetic methods presently available for analyzing genetic influences in insect host-parasitoid relationships. Most of the examples presented are taken from investigations with populations of *Drosophila melanogaster* infected with the larval parasitoid *Leptopilina boulardi* (Barbotin *et al.*, 1979; Nordlander, 1980; Carton *et al.*, 1986). Unquestionably, this is the most suitable natural and experimental host-parasitoid system available for investigating the genetic aspects of both host-immune competence and parasitoid virulence. However, will be seen, several of the genetic methods reported here are now sufficiently well documented to permit genetic studies of the immune reaction at the Mendelian and populational levels in other insect species.

Early Studies on Genetic Aspects of Immune Reaction

Early studies of resistance in insect host populations exposed to parasitoids introduced for purposes of biological control, demonstrated considerable variation in the defence responses of different populations. Studies by Vey and Vago (1969) suggested a genetic basis for host encapsulation responses in *Galleria* larvae exposed to the entomopathogen fungus *Aspergillus* spores. Resistant genotypes perhaps characterized those individuals that survived infection by producing a strong haemocytic response that resulted in the melanization and encapsulation of the fungal hyphae. Observations of Muldrew (1953) on the development of resistance of a tenthredinid species against a parasitoid species introduced for biological control can now be explained as a process of natural selection of a trait genetically determined. Salt and Van Den Bosch (1967) have demonstrated divergences in encapsulation of the same parasitoid amongst three host populations of a species located in California. These population differences certainly have a genetic basis. Other notable examples of the influence of genetic factors in insect host immune responses include resistant and refractory strains of the mosquito (Christensen, 1986). Collins *et al.* (1986) were able to produce fully refractory and fully susceptible strains of *Anopheles* through a short series of selective steps. However, studies of the *Drosophila-Leptopilina* interactions have provided the most convincing documentation to date to support the genetic basis for insect host immunity and the mechanisms of immune evasion employed by their parasites to avoid destruction. The very comprehensive studies by Schlegel-Oprecht (1953), Walker (1959) and Hadorn and Walker (1960) provided some of the first evidence for the genetic basis of host immunity in various natural populations of *Drosophila* and evasion of host defences by various parasitoid strains. Crosses between two natural populations with different capacities to encapsulate *Leptopilina* seem to indicate that the resistant character could be determined by a polyfactorial system with an incomplete dominance since hybrids exhibit some degree of response intermediate between the two parental strains. More recent supporting studies have been done using genetic mutants which either form melanotic tumours that resemble encapsulation, or certain enzyme-deficient mutants which cannot produce melanotic manifestations (Rizki and Rizki, 1980, 1983, 1990; Nappi and Silvers, 1984).

Utilization of the Isofemale Line Method

A relatively new approach to evaluating the genetic variability in natural populations is the isofemale line method. An isofemale line is set up for laboratory culture from a single inseminated founder female collected in the field (Parsons, 1980). The wild inseminated females can be regarded as 'pairs' if mated once or as 'families' if mated with several males. Such a line is a population that has undergone a single bottleneck. We employed this method to investigate a natural population of *D. melanogaster* collected from Brazzaville (Republic of Congo). In this locality, the parasitoid *L. boulardi* elicits a very efficient defence reaction from this host (Carton *et al.*, 1986; Carton, 1988a; Rizki *et al.*, 1990; Carton and Nappi, 1991). Since 1980, 22 isofemale lines established from this population have been kept in separate laboratory cultures.

Several isofemale lines are advantageous because selection and drift tend to cause loss of genetic variability in a single laboratory colony. Further, there is a persistence over many generations of consistent line differences originating from the founder effect. Parsons (1983) considered that after 10 generations (with $n = 100$ at each generation) 74% of the variance remains in the line. Although each isofemale line is separately highly inbred and homozygous, mixed together again, this new stock is genetically more representative of the field population than a single culture maintained in more or less good conditions. Delpuech *et al.* (1991) observed that in a mass culture started from 30 lines (5 females per line) kept during 23 generations under laboratory conditions, allele frequencies at three loci were similar to those observed initially in the field population.

Thus by maintaining numerous lines in laboratory cultures, the researcher can maximize the amount of genetic variability retained from the natural populations. Genetic investigation can also be performed with this technique for those characters whose genetic determinism is not known, permitting the assessment of genotypic variation at the populational level. Isofemale lines could also be the first step in a screening process of genotypes with opposite responses. With such selected stocks, it is possible to perform some precise investigtions on the genetic inheritance of the trait selected.

This method enabled us to maintain over several years the capacity of the *D. melanogaster* population to encapsulate. Thus, in 1989, we created again a mass culture starting from 5 females per line (i.e., 110 females) from the set of isofemale lines established in 1980. Testing this culture, we obtained a level of encapsulation of about 61.9%, similar to the value observed (55.7%) just after the capture. This constitutes evidence that with isofemale line breeding, the gene(s) of resistance were not lost and remained at the same frequency as for enzymatic alleles during the nine years of isofemale line culture.

Another advantage of the isofemale line method is that genetic studies can be done on a single generation by comparing the variances within and between lines. A significantly higher variability between lines is usually considered as the result of genetic divergences between lines. However, such a result is not an absolute proof of a genetic variability, especially when physiological traits are considered. Such traits are extremely sensitive to minor environmental variations, which are generally not identified and thus not controlled: significant differences can be due to the fact that the various lines are, in some respect, raised under different environmental conditions since they are bred in separate vials.

To ascertain the genetic component of variability, it is necessary not only to observe significant differences between lines, but also to obtain a good correlation between the measurements in the same line when successive generations are compared. Such an investigation was performed on the African population of *D. melanogaster* (Carton and Bouletreau, 1985). Depending on the line, the rate of encapsulation varied between 11.9% and 92.7% at the 8th generation reared in the laboratory and between 11.1% and 100% at the 15th generation, i.e., about 3 months later. This analysis enabled us to calculate the correlation between the two generations. This correlation was determined to be significant ($r = 0.62$, $df = 21$), indicating that the encapsulation capacity in *Drosophila* was partially or totally genetically determined.

An analysis of variance permits calculation of the within (Vw) and between (Vb) components of the total variance ($Vt = Vw + Vb$). From these components it is then possible to estimate the intraclass correlation ($t = Vb/Vt$). This later parameter measures the average likeness of individuals belonging to the same group (Falconer, 1981), i.e., individuals belonging to isofemale lines. Hoffmann and Parsons (1988) termed the intraclass correlation estimate obtained from the isofemale line procedure, the 'isofemale heritability'. As for the 'populational heritability' defined by Slatkin (1981), it is intermediate between heritability in a narrow sense and heritability in a broad sense. The confidence interval of this heritability depends on the confidence interval of the intraclass correlation. The following estimation of variance can be used:

$$var(t) = \frac{2(1-t)^2 \ (1+(n-1)t)^2}{n(n-1)(N-1)}$$

where n is the number of individuals measured per line and N the number of lines (Bulmer, 1985; Donner and Wells, 1986).

For the African population studied, we calculated an 'isofemale heritability' of encapsulation capacity of 0.43 with a variance of 0.14 (for n = 2 and N = 22).

The correlation between successive generations might provide another approach to the heritability of a trait. In our case, we only deal with the means of each isofemale line. Therefore, it was not possible to obtain an accurate kinship between two individuals belonging to the same line and/or individuals of different generations. However, as stressed by Capy (1987), if the effective size of each line is not too small, the correlation between successive generations can be used as an estimate of the upper limit of heritability in the narrow sense. In the case of the African population, the constancy of the cellular encapsulation capacity in each line over a short period (3 months) was tested by correlation analysis. As mentioned earlier, the values at the 8th generation and the 15th generation correlated very well (r = 0.62; P < 0.01). This value could represent the upper value of heritability.

The isofemale line method can also be used to screen and select inbred strains with opposite responses. Directional selection with some chance of success can also be based upon a combination of extreme lines from a set of isofemale lines. It is possible in this way to improve the efficiency of the selected trait in some stocks on the one hand, and to obtain stocks with no reaction on the other.

To develop our biological model, we used this approach during selection of a highly reactive strain by breeding for two successive generations, individuals which had issued from the mass culture created in 1989 that exhibited an encapsulation capsule in their abdomen after infection (Carton *et al.*, 1986). The non-reactive strain was obtained from the isofemale line whose progeny showed the lowest immune capacity (Carton and Bouletreau, 1985); after nine years of laboratory culture the level of encapsulation in this susceptible line was only about 2%, as had been predicted by the genetic drift effect.

Diallele Cross Technique

This technique, much more time-consuming, enables estimation of the additive genetic variance and the dominance variance, from which heritabilities in both the narrow and broad sense can then be calculated (for more details, see Crusio *et al.*, 1984). The technique consists of all possible matings between several inbred strains with different responses. For example, for 3 strains, there are 9 combinations (AA, BB, CC, AB, AC, BC and BA, CA and CB according to the sexes of the parents). Only isogenic strains or at least highly homozygous ones are suitable in implementing this technique. Full-sib crosses over several generations will produce the requisite homozygous strains.

Method to Evaluate Hereditary Chromosomal and Non-Chromosomal Components

Having constructed an immune reactive (R) line and maintained a susceptible (S) line of *D. melanogaster*, we recently began to analyze in detail the genetic mechanism influencing these two opposite immune states, employing the methods of Wahlsten (1979) and de Belle and Sokolowski (1987). Two generations of reciprocal crosses were made between the R and S lines to give 16 stocks of progeny: the two parental strains (P_1 and P_2), two F_1 hybrids, 8 back-crosses (4 each of the two well-defined phenotypes, R and S) and 4 F_2 hybrids (Table 1). Several replicates were made of each type of cross and each line was examined for encapsulation capacity. The following comparisons were done to determine the contribution of chromosomal and/or cytoplasmic components and the influence of sex on the immune capacities of the various types of hosts. The crosses are given in parentheses and the numbers refer to those described in Table 1.

Table 1: 16 crosses between reactive (R) strain and sensitive (S) strain used to separate their hereditary components

Cross No.	Mother	×	Father
Parental Strains			
1	S	×	S
2	R	×	R
Reciprocal F_1 Hybrids			
3	S	×	R
4	R	×	S
Reciprocal Back-crosses			
5	S	×	(S × R)
6	S	×	(R × S)
7	R	×	(S × R)
8	R	×	(R × S)
9	(S × R)	×	S
10	(R × S)	×	S
11	(S × R)	×	R
12	(R × S)	×	R
Reciprocal F_2 Hybrids			
13	(S × R)	×	(S × R)
14	(S × R)	×	(R × S)
15	(R × S)	×	(S × R)
16	(R × S)	×	(R × S)

1. S versus R parental strains (1 vs 2) to test the differences between the two parental lines.
2. S + R versus F_1 to investigate complete genetic dominance or additive effect (1 + 2 vs 3 + 4).
3. F_1s for deviation from an autosomal mode of inheritance (3 vs 4), i.e., non-autosomal inheritance (sex chromosomes, permanent cytoplasmic factors, transient maternal factors.
4. F_2s for significance of permanent cytoplasmic factors (13 + 14 vs 15 + 16).
5. Back-crosses to females (5 + 8 vs 6 + 7) to examine interaction Y chromosome and/or X chromosome and all other factors.
6. Back-crosses to males (9 + 12 vs 10 + 11) to examine interactions between the permanent cytoplasm and all other factors.

Statistical tests must utilize the contrast analysis of variance (Anova), which can be performed using the S.A.S. general linear models procedure (S.A.S. Institute Inc., 1985). If the character studied is expressed by a percentage, the statistical tests must be performed on arc sine transformed values.

With the *Drosophila* model, comprised of R and S lines, preliminary results seemed to indicate the presence of a complete dominance effect and that non-autosomal inheritance (sex chromosomes, cytoplasmic factors and transient maternal factors) was not involved.

A more complete statistical investigation than comparison 2 (see above) was then performed to determine whether experimental cross results fitted better a complete autosomal dominance pattern than a strictly additive pattern of inheritance. The following relationship would be present in the case of R totally dominant over S:

$$S \; < \; B_S \; < \; F_2 \; < \; (F_1 \; = \; B_R \; = \; R)$$

With a strictly additive pattern, supposing the effects of resistant and sensitive alleles equal but opposite, B_S should be intermediate between the S parental strain and the F_1 since it is comprised of 50% S/S and 50% R/S. The F_2 should not differ significantly from the F_1 since it is comprised of 25% S/S, 50% S/R and 25% R/R. B_R should be intermediate between the F_1 and the R/R parental strain since it is comprised of 50% R/S and 50% R/R. We obtained the following relationship in the case of a polygenic character with additive effect:

$$S \; < \; B_S \; < \; (F_2 \; = \; F_1) \; < \; B_R \; < \; R$$

Mendelian Analysis

The hypothesis of a simple genetic model was suggested by the preceding statistical investigations, which revealed the involvement of only chromosomal components. Under these conditions, observed ratios of reactive (R)/non-reactive or sensitive (S) individuals can be compared to expected ratios and evaluated by assuming a one-gene, complete dominance model.

In this model, expected ratios (E.R), representing the ratio number of reactive individuals to the number of sensitive individuals (or R : S), would be the following for each cross:

S × S	E.R. = 0 : 1
R × R	E.R. = 1 : 0
F_1	E.R. = 1 : 0
B_S	E.R. = 1 : 1
B_R	E.R. = 1 : 0
F_2	E.R. = 3 : 1

A chi-square 2 × 2 analysis will determine whether the observed data fit a single-gene, complete dominance model of inheritance of the relevant traits.

Method to Evaluate the Number of Genes Involved

In the case of cross results which better fit an additive effect model, we may suppose that the trait has a polygenic determinism. The number of independent genes with additive effects that contribute to the expression of a quantitative trait can be evaluated and is given by the Wright formula (Castle, 1921; Raymond *et al.*, 1987):

$$N = (U_1 - U_2)^2/(8 * \sigma^2)$$

(U_1 and U_2 refer to the character in the two parental strains; σ^2 = genetic variance).

Lande (1981) proposed various estimations of genetic variance, depending on the cross under consideration. Such values may underestimate the actual number of genes involved if they are linked or if they have no truly additive effect. The following values of genetic variance were utilized:

$$\sigma^2 = \sigma^2_{F_2} - \sigma^2_{F_1}$$

$$\sigma^2 = \sigma^2_{F_2} - (1/2 \ \sigma^2_{F_1} + 1/4 \ \sigma^2_R + 1/4 \ \sigma^2_S)$$

$$\sigma^2 = 2 \ \sigma^2_{F_2} - \sigma^2_{B_R} - \sigma^2_{B_S}$$

$$\sigma^2 = \sigma^2_{B_R} + \sigma^2_{B_S} - (\sigma^2_{F_1} + 1/2 \ \sigma^2_R + 1/2 \ \sigma^2_S)$$

Conclusion

Insect-borne parasitic diseases such as malaria, trypanosomiasis, leishmaniasis and filariasis pose very serious health problems throughout the world. Identifying the factors that influence the success or failure of these insect host-parasite associations would contribute significantly to an understanding of the epidemiology of these diseases. One important type of immune response made by insects against the metazoan parasites they transmit involves the blood cells or haemocytes in the formation of a melanotic capsule around the foreign body. Cell-free or humoral mechanisms are employed by insects in their defence against certain bacterial, viral and protozoan parasites. As effective as these defences are in maintaining self-integrity, many parasites successfully circumvent them by cellular and biochemical methods (Rizki and Rizki, 1984, 1990; Carton *et al.*, 1989) that rival those of their hosts for effectiveness and specificity. Genetic reciprocity is the basis for the adaptive strategies developed by these combatants (Bouletreau, 1986; Carton and Nappi, 1991) in their co-evolutionary struggles. Knowledge of the interacting genetic mechanisms which govern host immunity and parasite resistance is fundamental to developing a comprehensive understanding of the physiological and biochemical aspects of these responses. Additionally, since these co-evolutionary adaptations operate at the populational level, knowledge of the genetic components involved in insect host-parasite associations will impact significantly on the development of preventive measures and determination of how these efforts can be most effectively applied to natural populations. The *Drosophila* host-parasite association represents a suitable model for investigating the genetics of the immune response; however, such could also be investigated in other insect species, undoubtedly with some success.

REFERENCES

Barbotin, F., Y. Carton and S. Kelner-Pillault. 1979. Morphologie et biologie de *Cothonaspis (Cothonaspis) boulardi* n. sp., parasites des drosophiles. *Bull. Soc. Ent. Fr.* 84: 19.

Bouletreau, M. 1986. The genetic and coevolutionary interactions between parasitoids and their hosts. In: *Insect Parasitocids* (J. Waage and D. Greathead, eds.), pp. 169-200, Academic Press, London.

Bouletreau, M. and P. Fouillet. 1982. Variabilité génétique intrapopulation de l'aptitude de *Drosophila melanogaster* à permettre le développement d'un Hyménoptère parasite. *C.R. Acad. Sci.* (Paris) 295: 775-778.

Bulmer, M.G. 1985. *The Mathematical Theory of Quantitative Genetics*, Clarendon Press, Oxford.

Capy, P. 1987. Variabilité génétique des populations naturelles de *Drosophila melanogaster* et de *Drosophila simulans*. Thése de Doctorat, Université de Paris.

Carton, Y. 1988a. Ecological and genetic significance of the encapsulation process in *Drosophila* against a parasitic wasp. In: *3rd Eur. Workshop on Parasitoids of Insects*, 123-125 (M. Bouletreau and G. Bonnot, eds.). INRA.

Carton, Y. 1988b. La Coévolution. *La Recherche* 202: 1021-1031.

Carton, Y. and M. Bouletreau. 1985. Encapsulation ability of *D. melanogaster*: A genetic analysis. *Dev. Comp. Immunol.* 9: 211-219.

Carton, Y. and A. Nappi. 1991. The *Drosophila* immune reaction and the parasitoid capacity to evade it: Genetic and coevolutionary aspects. *Acta Oecol., J. Int. Ecol.* 12: 89-104.

Carton, Y., P. Capy and A. Nappi. 1989. Genetic variability of host-parasite relationship traits: Utilization of isofemale lines in a *Drosophila* parasitic wasp. *Genetics, Selection, Evolution* 21: 437-446.

Carton, Y., M. Bouletreau, J. Van Alphen and J. Van Lenteren. 1986. The *Drosophila* parasitic wasps. In: *The Genetics and Biology of Drosophila*, pp. 347-394 (M. Ashburner, L. Carson and J.N. Thompson S.T. eds.). Academic Press, New York.

Castle, W.E. 1921. An improved method of estimating the number of genetic factors concerned in cases of blending inheritance. *Science* 54: 223-225.

Christensen, B.M., 1986. Immune mechanisms and mosquito-filarial worm relationships. In: *Zool. Soc. London Symposia*, 566, pp. 145-160 (A.M. Lackie, ed.). Clarendon Press, Oxford.

Collins, F.H., R.K. Sakai, K.D. Vernick, S. Paskewitz, D.C. Seeley, L.H. Miller, L.H. Collins, C.C. Campbell and R.W. Gwadz. 1986. Genetic selection of a plasmodium refractory strain of the malaria vector *Anopheles gambiae*. *Science* 234: 607-610.

Crusio, W.E., J.M.L. Kerbusch and J.H.F. van Abeelen. 1984. The replicted diallel cross: a generalized method of analysis. *Behavior Genetics* 14: 81-104.

de Belle, J.S. and M. Sokolowski. 1987. Heredity of rover/sitter: Alternative foraging strategies of *Drosophila melanogaster* larvae. *Heredity* 59: 73-83.

Delpuech, J.M., Y. Carton and R.T. Roush. 1991. How to conserve genetic variability of a wild insect population in laboratory conditions. (in litt.).

Donner, A. and G. Wells. 1986. A comparison of confidence interval methods for the intraclass correlation coefficient. *Biometrics* 42: 401-402.

Falconer, D.S. 1981. *Introduction to Quantitative Genetics*. Longman, London.

Götz, P. and H.G. Boman. 1985. Insect Immunity. In: *Comprehensive Insect Physiology, Biochemistry and Pharmacology*, pp. 453-485 (G.A. Kerkut and L.I. Gilbert, eds.). Pergamon Press, Oxford.

Hadorn, E. and I. Walker. 1960. *Drosophila* and *Pseudeucoila*. I. Selektionsversuche zur steigerung der Abwehrrraktion des Wirtes gegen den Parasiten. *Revue Suisse de Zoologie* 67: 216-225.

Hoffmann, A.A. and P.A. Parsons. 1988. The analysis of quantitative variation in natural populations with isofemale strains. *Gen. Sel. Evol.* 20: 87-98.

Lande, R. 1981. The minimum number of genes contributing to quantitative variation between and within populations. *Genetics* 99: 541-553.

Muldrew, J.A. 1953. The natural immunity of the larch sawfly (*Pritisphora erichsonii* Htg.) to the introduced parasite *Mesoleius tenthredinis* Morley in Manitoba and Saskatchewan. *Can. J. Zool.* 31: 313-332.

Nappi, A. and M. Silvers. 1984. Cell surface changes associated with cellular immune reactions in *Drosophila*. *Science* 225: 1166-1168.

Nappi, A. and Y. Carton. 1986. Cellular immune responses of *Drosophila*. In: *Immunity in Invertebrates*, 13: 171-187 (M. Brehelin, ed.). Springer-Verlag, Berlin-Heidelberg.

Nordlander, G. 1980. Revision of the genus *Leptopilina* Forster, 1939, with notes on the status of some other genera (Hymenoptera, Cynipoidae: Eucoilidae). *Entomol. Scand.* 11: 428-453.

Parsons, P.A. 1980. Isofemale strains and evolutionary strategies in natural populations. *Evol. Biol.*, 13: 175-217.

Parsons, P.A. 1983. *The Evolutionary Biology of Colonizing Species*. Cambridge University Press, Cambridge.

Raymond, M., N. Pasteur and G.P. Georghiou. 1987. Inheritance of chlorpyrifos resistance in *Culex pipiens* (Diptera: Culicidae) and estimation of the number of genes involved. *Heredity* 58: 351-356.

Rizki, T.M. and R.M. Rizki. 1980. Development analysis of a temperature-sensitive melanotic tumor mutant in *Drosophila melanogaster*. *Wilhem Roux's Arch. Dev. Biol.* 189: 197-206.

Rizki, T.M. and R.M. Rizki. 1983. Blood cell surface changes in *Drosophila* mutants with melanotic tumors. *Science* 220: 73-75.

Rizki, T.M. and R.M. Rizki. 1984. Selective destruction of a host blood cell type by a parasitoid wasp. *Proc. Natl. Acad. Sci.* 81: 6154-6158.

Rizki, T.M. and R.M. Rizki. 1990. Encapsulation of parasitoid eggs in phenoloxidase-deficient mutants of *Drosophila melanogaster*. *J. Insect Physiol.* 36: 5239.

Rizki, T.M., R.M. Rizki and Y. Carton. 1990. *Leptopilina heterotoma* and *L. boulardi*: Strategies to avoid cellular defense responses of *Drosophila melanogaster*. *Exper. Parasit.* 70: 466-475.

Salt, G. and R. van den Bosch. 1967. The defence reactions of three species of *Hypera* (Coleoptera, Curculionidae) to an ichneumon wasp. *J. Invert. Pathol.* 9: 164-167.

Schlegel-Oprecht E. 1953. Versuche zur Auslsung von Mutationen bei der Zoophagen Cynipide *Pseudeucoila bochei* Weld und Befunde über die Stammspezifische Abwehrreaktion des Wirtes *Drosophila melanogaster*. *Zeit. f. Ind. Abst. u. Vererbungslehre* 85: 245-281.

Slatkin, M. 1981. Fixation probabilities and fixation times in a subdivided population. *Evolution* 35: 477-488.

Stoffolano, J.G. 1986. Nematode-induced host responses. In: *Hemocytic and Humoral Immunity in Arthropods*, pp. 117-156. (A.P. Gupta, ed.). John Wiley and Sons, New York.

Vey, A. and C. Vago. 1969. Recherches sur la guérison dans les infections cryptogamiques d'Invertébrés: infection à *Aspergillus niger* v. Tiegh. chez *Galleria mellonella* L. *Ann. Zool. Ecol. Anim.* 1: 121-126.

Wahlsten, D. 1979. A critique of the concepts of heritability and heredity in behavioral genetics. In: *Theoretical Advances in Behavioral Genetics*, pp. 426-481 (J.R. Royce and L. Mos, eds.). Sijthoff and Nordhoff, Germantown, MD.

Walker, I. 1959. Die Abwehrreaktion des Wirtes *Drosophila melanogaster* gegen die zoophage Cynipidae *Pseudeucoila bochei* Weld. *Rev. Suisse Zool.* 66: 569-632.

CHAPTER 8

Interactions between the Insect Endocrine System and the Immune System

S. Bradleigh Vinson

Introduction

Insects have a chitinous exoskeleton which they must periodically shed (moult) in order to grow. This process, which includes the partial digestion of the old exoskeleton, formation of the new, shedding of the remaining old exoskeleton, expansion and hardening of the new exoskeleton, is under the control of the insects neuroendocrine system. This system involves three major organs and the hormones released by each. The process begins when neurosecretory cells located in the insect's brain release in the haemolymph a neuropeptide referred to as prothoracicotropic hormone (PTTH). The prothoracic gland is stimulated by PTTH to produce and release a steroid hormone, α-ecdysone. The α-ecdysone is converted to the active beta form by several insect tissues and initiates the moulting process.

Along with the release of ecdysone by the prothoracic gland, the corpora allata releases a terpenoid (juvenile hormone) that maintains the insect's immature tissues and suppresses adult tissue expression and development. The presence of both juvenile hormone and β-ecdysone results in a larval to larval moult (Wigglesworth, 1964). In the absence of the juvenile hormone, the adult tissues are expressed and the larval tissues degrade during the moult (Steele and Davey, 1985). In the adult, juvenile hormones are again secreted and affect many adult functions.

These two hormones, the juvenile hormone (JH) and ecdysone, not only regulate the moulting process, but appear to affect many other systems (Steele, 1985; Hagedorn, 1985; Denlinger, 1985; Hardie and Lees, 1985). For example, ecdysone has been reported to influence melanization (Chang and Jang, 1980), protein storage in the fat body (Dean et al., 1980; Locke,

1980), muscle degeneration (Zachary and Hoffmann, 1980) and Malpighian tubule function (Ryerse 1986), while the juvenile hormone influences reproduction (Doane, 1973), accessory gland function (Weaver, 1981) and fat body metabolism (Chess and Wyatt, 1981).

Although the juvenile hormone and ecdysone exert an effect on many physiological systems (Chess and Wyatt, 1981), their affect on the insect immune system and the insect's immune response is not well documented. As described in other chapters of this book, the immune system of insects involves both humoral as well as cellular components. The humoral components are either involved in recognition or are directly responsible for the immune response. These humoral components include the bactericidal proteins (Boman, 1981; Boman and Hultmark, 1981; Dunn *et al.*, 1985), the lectins (Olafsen, 1986; Rowley *et al.*, 1986; Renwarantlz, 1986), mucoproteins (Anderson and Chain, 1986; Chain and Anderson, 1983), melanin and the phenoloxidase system (Söderhäll and Ajaxon, 1982, Söderhäll and Smith, 1986) and lysozymes (Stebbins and Hapner, 1986; Hapner and Stebbins, 1986). The cellular component primarily involves the haemocytes, which occur in a variety of forms and have a variety of functions (Rowley and Ratcliffe, 1981; Brehelin and Zachary, 1986). Among the various forms of haemocytes, the plasmatocytes and granulocytes or 'similar' cells have most often been implicated in the cellular immune response (Gupta, 1985; Lackie, 1988). Of particular importance are the phagocytic cells, which engulf or encapsulate foreign material (Götz and Boman, 1985; Ratcliffe and Rowley, 1979) and remove it from the system. Aside from the role of the bactericidal proteins (Dunn, 1986) and the phagocytic cells (Gupta, 1985; Götz and Boman, 1985), little is known of the function of the other factors in the immune system.

Haemocytes and the Endocrine System

Haemocytes appear to be important to the growth and moulting of insects. When the haemocytes are damaged, as occurs with an injection of a high amount of trypan blue, iron saccharate or India ink (Wigglesworth, 1955a), there is a delay in the moult of *Rhodnius prolixus* by 1-3 weeks, if the haemocytes were damaged early (1-3 days) in the immature insect feeding stage. Further, when the haemocytes are damaged, the effect on the insects moult is more sharply defined than with decapitation or abdomen isolation and occurs slightly later (Wigglesworth, 1955b). Such results led Wigglesworth (1979) to suggest that haemocytes transport ecdysone to the epidermis. However, which haemocytes may be involved in such transport is unknown, albeit there is evidence that carrier proteins exist (Smith, 1985; Goodman and Chang, 1985).

Endocrine Effects of the Immune System

Whether the state of the neuroendocrine system or hormones has an effect on the insect immune system has not been established nor has it received much attention. A number of reports indicate that young insects (the earlier instars) are less effective in encapsulating foreign objects than later instars (Van der Bosch, 1964; Lynn and Vinson, 1977; Brewer, 1971; Berberet, 1986; Blumberg, 1977, 1982; Blumberg and DeBach, 1981; Schneider, 1950; Puttler, 1967). For example, Nappi and Streams (1969) studied the immune response of *Drosophila* to the cynipid parasitoid *Leptophilina heterotoma* (= *Pseudeucoila bochei*) and reported that parasitoid success was greater in younger than in older hosts. They concluded that immunity appeared to be a function of age. Salt (1963) attributed the increased percentage encapsulation of introduced foreign material in later instars to the larger number of available haemocytes. Van Driesche *et al.* (1986), however, reported that encapsulation of the parasitoid *Epidinocaris diveisicornis* in older mealybugs, *Phenacoccus herreni*, was lower than in 2nd instar nymphs. Whether the older mealybugs have fewer haemocytes is not reported, but frequently the haemocyte numbers in adult insects are less than in the immature (Arnold, 1974).

Haemocyte counts often vary with the treatment of insects and their physiological condition (Yeager, 1945; Jones, 1967; Gupta, 1985; Wheeler, 1963) but whether these changes are significant in the host's ability to respond immune-competently is unknown. Haemocyte numbers per unit volume often increase in immature insects from one instar to the next (Arnold and Hinks, 1976; Wago and Ishikawa, 1979; Dumphy and Nolan, 1980; Webley, 1951). Yet, during the period from one ecdysis to the next, there is a slight decrease in the total number of haemocytes per unit volume (Patton and Flint, 1959; Pathak, 1986; Lee, 1961; Brengnon and Le Berre, 1976), particularly just before ecdysis. The result is a fluctuation in the slow increase in haemocyte numbers among the developing immatures. Decreases have been most often recorded from pupae (Pelc, 1986; Walters, 1970; Arnold, 1952) with numbers again increasing in the young adult (Walters, 1970), but not reaching the level recorded in the immature (Arnold, 1974). There are also changes in the haemocyte population in regard to types during the development of many species (Gupta, 1985; Götz and Boman, 1985; Peake, 1979; Wago and Ishikawa, 1979).

Changes in the composition of the population of haemocytes during development resemble, at least in Diptera, those changes observed during encapsulation (Stoffolano and Streams, 1971; Walker, 1959; Nappi and Stoffolano, 1971, 1972; Nappi and Streams, 1969). Such changes include: haemocyte mobilization, differentiation, and even lysis of certain cells (Nappi, 1974). In *Musca domestica*, parasitization by the nematode *Heterotylenchus autumnalis* resulted in an increase in the total haemocyte count during encapsualtion (Nappi and Stoffolano, 1971). In another dipteran, *Orthellia caesarion*, the same invading nematode was only encapsulated in adults (Nappi and Streams 1971). Vey (1971) concluded that the response of Lepidoptera to the fungi, *Aspergillus*, was by the aggregation of haemocytes followed by their melanization, but the melanin reaction was more intense in pupae. Thus, there is no doubt that the response of hosts to a pathogen can be related to host stage. During development of *Drosophila melanogaster* there is an increase in the transformation of plasmatocytes to lamellocytes in late-instar larvae and pupae, while the crystal cells (which resemble oenocytoids) disappear in pupae (Rizki, 1957; Rizki and Rizki, 1959; Nappi and Streams, 1969). In *D. melanogaster* parasitized by *L. heterotoma*, when the parasitoid was successful in development, there was a slight increase in the number of crystal cells and a decrease in the number of lamellocytes (Nappi and Streams, 1969)—a reverse of the results observed in healthy insects. In contrast, in *D. melanogaster* resistant to *L. heterotoma*, the crystal cells prematurely decreased while there was a precocious transformation of plasmatocytes to lamellocytes (Walker, 1959). But, are such changes in total or differential haemocyte counts influenced by the hormonal system? Rizki (1957, 1960, 1962) examined the effects of isolation of the endocrine system on haemocyte transformation and tumourigenesis in *Drosophila melanogaster*. Larvae were ligated to isolate haemocytes from the hormonal system located anterior to the ligation. Rizki (1960, 1962) found a precocious transformation of plasmatocytes to lamellocytes in ligated animals and that the process could be halted by endocrine gland implants.

Working with a true bug, *Halys dentata*, Pathak (1983) also examined the effects of extirpation and implantation of the brain, corpora allata and corpora cardiaca on the total haemocyte count. When the corpora allata and corpora cardiaca were extirpated, the haemocyte counts were significantly depressed for over ten days. Extirpation of the corpora cardiaca alone, however, only resulted in a decrease in haemocyte numbers after six days, while extirpation of the corpora allata alone resulted in only a temporary 2-5-day depression of haemocyte numbers. The reverse occurred through implantation. Implanted corpora allata increased the total haemocyte numbers for 4-6 days. Implantation of young corpora cardiaca (one-day-old adult) into one-day-old adult bugs had no effect, while implantation of six-day-old adult corpora cardiaca into similar one-day-old adult bugs increased the total haemocyte count (Pathak, 1983). The reasons for these changes are not clear, but the results again indicate that hormones might exert some effect on haemocyte numbers. Jones and Liu (1969) through ligation experiments using *Galleria mellonella* also suggested that the endocrine system might have an effect on the production and release of haemocytes. A similar conclusion was reached by Hoffmann (1970) using *Locusta migratoria*. But, as noted by Feir (1979) in her review, direct evidence that the hormone system affects the production of haemocytes is not established.

These results suggest that the hormone system might play a role in haemocyte transformation. Based on such results, Nappi (1975) examined the effect of ligation of *D. algonquin*, a host that readily encapsulates

L. heterotoma, on the encapsulation process. He found a decrease in the encapsulation rate of parasites posterior to the ligation, supporting the proposition that the cell-mediated response results from a hormonal imbalance (Nappi 1973a and b, 1975). Nappi (1974) further suggested that low JH, which occurs during pupation, might bring about the precocious haemocyte changes observed after parasitizm and during successful encapsulation. Juvenile hormone has been reported to increase tumourigenesis (Bryant and Sang, 1969; Madhaven, 1972), which is accompanied by haemocyte changes similar (Rizki, 1957, 1960) to those in parasitizm (Nappi 1973a and b).

Hormone Effects on the Immune Response

There have been relatively few studies on the effects of any of the hormones, for example ecdysone or juvenile hormone, on haemocytes or the immune response. Das and Gupta (1977) reported that the injection of a juvenile hormone analogue into *Blatella germanica* altered the total and differential haemocyte count, which resulted in a decrease in plasmatocytes, but no change in granulocytes was observed. A 40% increase in coagulocytes, which resulted in melanization of the JH-treated adultoids, also occurred (Das and Gupta, 1977). Whether the JH altered the ability of the haemolymph to melanize was not reported. However, the plasmatocytes appeared to degenerate when exposed to excessive juvenile hormone (Gupta, 1986).

Haemocytes exist as monodispersed cells but during the formation of capsules around foreign objects, a process unique to arthropods (Baerwald, 1979), the haemocytes come into contact with each other. One aspect of these capsules, which are composed of layers of cells, is the occurrence of many gap junctions (Grimstone *et al.*, 1967; Norton and Vinson, 1978). The frequency of gap junctions has been reported to be influenced by ecdysone in *Limulus* (Johnson *et al.*, 1974), but no studies were found in the literature on the effects of hormones on the gap junctions of haemocytic capsules. Despite the possible influence of hormones on gap junction formulation, such an influence would not be expected to affect the initial capsule formation. However, gap junction formation might be important to the growth of a capsule. If so, then hormones might influence capsule thickness, although no evidence exists to support this supposition.

Changes in the relative number of phagocytic cells and their morphology following injection of crustecdysone into *Calliphora* was reported by Crossley (1968). However, such a response by the phagocyte could be due to a response from the degrading tissues, stimulated by the hormone injection, rather than a direct response to the hormone. Judy and Marks (1971) reported that haemocytes in *Manduca* increased their locomotory activity in response to ecdysone. But whether increased haemocyte mobility has any affect on the immune system is not known.

Lynn and Vinson (1977) examined the effects of injections of both a juvenile hormone analogue (hydroprene) and a mixture of α- and β-ecdysone on the encapsulation of both a foreign material (a cactus thorn) and the eggs of *Cardiochiles nigriceps* in two hosts (one permissive and one resistant to the parasitoid). The results did not suggest that there was any strong interaction between the hormones and the encapsulation response on the part of the host.

Conclusion

The evidence available at present suggests that differentiation and possibly haemocyte density, which is influenced by the haemolymph volume, may be influenced by components of the hormone system. Such changes in haemocyte numbers and the particular composition and degree of differentiation might influence the insect's immune response. However, such effects appear to be minor. No evidence of a major effect of the hormonal system on the insect immune system exists. The immune system of multicellular organisms is essential to the survival of these organisms. The PTTH-ecdysone-juvenile hormone system is involved in the discontinuous growth and development characteristic of arthropods. Thus, these hormones have ups and downs and their changing ratio influences differentiation (Jungreis, 1979). The organism cannot afford to respond to invading organisms on a discontinuous basis and thus such hromones as ecdysone, juvenile hormones or PTTH would not be expected to have any major influence on the insect immune system.

REFERENCES

Anderson, R.S. and B.M. Chain. 1986. Macrophage functions in insects. In: *Hemocytic and Humoral Immunity in Arthropods,* pp. 77-78 (A.P. Gupta, ed.). John Wiley and Sons, New York.

Arnold, J.W. 1952. The haemocytes of the Mediterranean flour moth, *Ephestia Kuhneilla* Zell. *Can. J. Zool.* 30: 352-364.

Arnold, J.W. 1974. The hemocytes of insects. In: *The Physiology of Insecta*, vol. 5, pp. 202-258 (M. Rockstein, ed.). Academic Press, New York.

Arnold, J.W. and C.F. Hinks. 1976. Hemopoiesis in Lepidoptera. I. The multiplication of circulating haemocytes. *Can. J. Zool.* 54: 1003-1012.

Baerwald, R.J. 1979. Fine structure of haemocyte membranes and intercellular junctions formed during haemocyte encapsulation. In: *Insect Hemocytes* pp. 231-258 (A.P. Gupta, ed.). Cambridge University Press, New York.

Berberet, R.C. 1986. Relationship of temperature of embryogenesis and encapsulation of eggs of *Bathyplectes curculionis* (Hymenoptera: Ichneumonidae) in larvae of *Hypera postica* (Coleoptera: Curculionidae). *Ann. Ent. Soc. Am.* 79: 985-988.

Blumberg, D. 1977. Encapsulation of parasitoid eggs in soft scales (Homoptera: Coccidae). *Ecol. Ent.* 2: 185-192.

Blumberg, D. 1982. Further studies of the encapsulation of *Metaphycus swirskii* by soft scales. *Entomologia Exp. Appl.* 31:·245-248.

Blumberg, D. and P. DeBach. 1981. Effects of temperature and host age upon the encapsulation of *Metaphycus stanleyi* and *Metaphycus helvolus* eggs by brown soft scale *Coccus hesperidum*. *J. Invert. Pathol.* 37: 73-79.

Boman, H.G. 1981. Insect responses to microbial infection. In: *Microbial Control of Insects, Mites, and Plant Diseases*, pp. 769-789 (D. Burges, ed.). Academic Press, New York.

Boman, H.G. and D. Hultmark. 1981. Cell-free immunity in insects. *Trends Biochem. Sci.* 6: 306-309.

Brehelin, M. and D. Zachary. 1986. Insect haemocytes: A new classification to rule out the controversy. In: *Immunity in Invertebrates*, pp. 36-48 (M. Brehelin, ed.). Springer-Verlag, Berlin.

Brengnon, M. and J. LeBerre. 1976. Variation de la formule hemocytaire et du volume d'hemolymphe chez la chenille de *Pieris brassicae* L. *Ann. Zool. Ecol. Anim.* 8: 1-12.

Brewer, R.H. 1971. The influence of the parasite *Comperiella bifasciata* How. on the populations of two species of armored scale insects. *Aonidiella aurantii* (Mask.) and *A. citrina* (Coq.), in South Australia. *Aust. J. Zool.* 19: 53-63.

Bryant, P.J. and J.H. Sang. 1969. Physiological genetics of melanotic tumors in *Drosophila melanogaster*. VI. The turmorigenic effects of juvenile hormone-like substances. *Genetics* 62: 321-336.

Chain, B.M. and R.S. Anderson. 1983. Inflammation in insects: The release of a plasmatocyte depletion factor following interaction between bacteria and haemocytes. *J. Insect. Physiol.* 29: 1-4.

Chang, F. and E. Jang. 1980. Endocrine regulation of melanization in late larvae of the oleander hawk moth, *Deilephila nerii*. *Ann. Entomol. Soc. Am.* 73: 89-92.

Chess, T.T. and G. Wyatt. 1981. Juvenile hormone control of vitellogenin synthesis in *Locusta migratoria*. In: *The Regulation of Insect Development and Behavior*, pp. 533-536 (F. Sehnal, A. Zabza, J.J. Menn and B. Cymborowski, eds.). Wroclaw Technical University Press, Wroclaw.

Crossley, A.C. 1968. The fine structure and mechanism of breakdown of larval intersegmental muscles in the blowfly, *Calliphora erythrocephola*. *J. Insect Physiol.* 14: 1389-1407.

Das, Y.T. and A.P. Gupta. 1977. Nature and precursors of juvenile hormone-induced excessive cuticular melanization in German cockroaches. *Nature* 268: 139-140.

Dean, R.L., W.E. Bollenbacher, M. Locke, L.I. Gilbert and S.L. Smith. 1980. Hemolymph ecdysteroid levels and cellular events in the intermolt/molt sequence of *Calpodes ethilus*. *J. Insect Physiol.* 26: 267-280.

Denlinger, D.L. 1985. Hormonal control of diapause. In: *Comprehensive Insect Physiology, Biochemistry and Pharmacology*, vol. 3. pp. 353-412 (G.A. Kerkut and L.I. Gilbert, eds.). Pergamon Press, New York.

Doane, W.W. 1973. Role of hormones in insect development. In: *Developmental Systems: Insects*, pp. 291-497 (S.J. Counce and C.H. Waddington, eds.). Academic Press, New York.

Dumphy, G.B. and R.A. Nolan. 1980. Hemograms of selected stages of the spruce budworm, *Choristoneura fumiferana*. *Canad. Entomol.* 112: 443-450.

Dunn, P.E. 1986. Biochemical aspects of insect immunology. *Annu. Rev. Entomol.* 31: 321-339.

Dunn, P.E., W. Dai, M.R. Kanost and C. Geng. 1985. Soluble peptidoglycan fragments stimulate antibacterial protein synthesis by fat body from larvae of *Manduca sexta*. *Dev. Comp. Immunol.* 9: 559-568.

Feir, D. 1979. Multiplication of haemocytes In: *Insect Haemocytes*, pp. 67-82 (A.P. Gupta, ed.). *Cambridge University Press*, Campridge.

Goodman, W.G. and E.S. Chang. 1985. Juvenile hormone cellular and hemolymph binding proteins. In: *Comprehensive Insect Physiology, Biochemistry and Pharmacology*, vol. 7, pp. 479-570 (G.A. Kerkut and L.I. Gilbert, eds.). Pergamon Press, New York.

Götz, P. and H. Boman. 1985. Insect immunity. In: *Comprehensive Insect Physiology, Biochemistry and Pharmacology*, vol. 3, pp. 453-485 (G.A. Kerkut and L.I. Gilbert, eds.). Pergamon Press, New York.

Grimstone, S.R., S. Rotheram and G. Salt. 1967. An electron-microscope study of capsule formation by insect blood cells. *J. Cell Sci.* 2: 281-292.

Gupta, A.P. 1985. Cellular elements in the hemolymph. In: *Comprehensive Insect Physiology Biochemistry and Pharmacology*, vol. 3, pp. 401-451 (G.A. Kerkut and L.I. Gilbert, eds.). Pergamon Press, New York.

Gupta, A.P. 1986. Arthropod immunocytes: Identification, structure, function, and analogies to the function of vertebrate B- and T-lymphocytes. In: *Haemocytic and Humoral Immunity in Arthropods*, pp. 3-59 (A.P. Gupta, ed.). John Wiley and Sons, New York.

Hagedorn, H.H. 1985. The role of ecdysteroids in reproduction. In: *Comprehensive Insect Physiology, Biochemistry and Pharmacology*, vol. 8, pp. 205-262 (G.A. Kerkut and L.I. Gilbert, eds.). Pergamon Press, New York.

Hapner, K.D. and M.R. Stebbins. 1986. Biochemistry of arthropod agglutinins. In: *Haemocytic and Humoral Immunity in Arthropods*, pp. 227-250 (A.P. Gupta, ed.). John Wiley and Sons, New York.

Hardie, J. and A.D. Lees. 1985. Endocrine control of polymorphism and polyphenism. In: *Comprehensive Insect Physiology, Biochemistry and Pharmacology*, vol. 8, pp. 441-490 (G.A. Kerkut and L.I. Gilbert, eds.). Pergamon Press, New York.

Hoffmann, J.A. 1970. Endocrine regulation of the production and differentiation of hemocytes in an orthopteran insect: *Locusta migratoria migratoroides*. *Gen. Comp. Endocr.* 15: 198-219.

Johnson, G., D. Quick, R. Johnson and W. Herman. 1974. Influences of hormones on gap junctions in horseshoe crabs. *J. Cell Biol.* 63: 157.

Jones, J.C. 1967. Effects of repeated hemolymph withdrawals and ligaturing the head on differential haemocyte counts of *Řhodnius prolixus* Stal. *J. Insect Physiol.* 13: 1351-1360.

Jones, J.C. and D.P. Liu. 1969. The effect of ligaturing *Galleria mellonella* larvae on total haemocyte counts and on mitotic indices among haemocytes. *J. Insect Physiol.* 15: 1703-1708.

Judy, K., and E.P. Marks. 1971. Effects of ecdysone *in vitro* on hindgut and haemocytes of *Manduca sexta*. *Gen. Comp. Endocr.* 17: 351-359.

Jungreis, A.M. 1979. Physiology of molting in insects. *Adv. Insect Physiol.* 14: 109-183.

Lackie, A.M. 1988. Hemocyte behavior. *Adv. Insect Physiol.* 21: 85-178.

Lee, R.M. 1961. The variation in blood volume with age in the desert locust (*Schistocerca gregaria* Forsk.). *J. Insect Physiol.* 6: 36-51.

Locke, M. 1980. The cell biology of fat body development. In: *Insect Biology in the Future*, pp. 227-252 (M. Locke and D.S. Smith, eds.). Academic Press, New York.

Lynn, D.C. and S.B. Vinson. 1977. Effects of temperature, host age and hormones upon the encapsulation of *Cardiochiles nigriceps* eggs by *Heliothis* spp. *J. Invert. Path.* 29: 50-55.

Madhaven, K. 1972. Induction of melanotic pseudotumors in *Drosophila melanogaster* by juvenile hormone. *Wilhelm Roux' Archiv.* 169: 345-349.

Nappi, A.J. 1973a. Hemocytic changes associated with the encapsulation and melanization of some insect parasites. *Exp. Parasit.* 33: 285-302.

Nappi, A.J. 1973b. The role of melanization in the immune reaction of larvae of *Drosophila algonquin* against *Pseudeucoila bochei*. *Parasitology* 66: 23-32.

Nappi, A.J. 1974. Insect hemocytes and the problem of host recognition of foreignness. In: *Contemporary Topics in Immunobiology*, vol. 4, pp. 207-224 (E.L. Cooper, ed.). Plenum Press, New York.

Nappi, A.J. 1975. Effects of ligation on the cellular immune reactions of *Drosophila algonquin* against the hymenopterous parasite *Pseudeocoila bochei*. *J. Parasitology* 61: 373-376.

Nappi, A.J. and F.A. Streams. 1969. Haemocytic reactions of *Drosophila melanogaster* to the parasites *Pseudeucoila mellipes* and *P. bochei*. *J. Insect Physiol.* 15: 1551-1566.

Nappi, A.J. and J.G. Stoffolano. 1971. *Heterotylenchus autumnalis*. Hemocytic reactions and capsule formation in the host, *Musca domestica*. *Exptl. Parasitology* 29: 116-125.

Nappi, A.J. and J.G. Stoffolano. 1972. Haemocytic changes associated with the immune reaction of nematode-infected larvae of *Orthellia caesarion*. *Parasitology* 65: 295-302.

Norton, W.N. and S.B. Vinson. 1978. Encapsulation of a parasitoid egg within its habitual host: An ultrastructural investigation. *J. Invert. Pathol.* 30: 55-67.

Olafsen, J.A. 1986. Invertebrate lectins—biochemical heterogeneity as a possible key to their biological function. In: *Immunity in Invertebrates: Cells, Molecules and Defence Reactions*, pp. 94-111 (M. Brehelin, ed.). Springer-Verlag, Berlin.

Pathak, J.P.N. 1983. Effect of endocrine glands on the unfixed total haemocyte counts of the bug *Halys dentata*. *J. Insect Physiol.* 29: 91-94.

Pathak, J.P.N. 1986. Hemogram and its endocrine control in insects. In: *Immunity in Invertebrates*, pp. 49-59 (M. Brehelin, ed.). Springer-Verlag, Berlin.

Patton, R.L. and R.A. Flint. 1959. The variation in blood cell count of *Periplaneta americana* (L.) during a molt. *Ann. Entomol. Soc. Amer.* 52: 240-242.

Peake, P.W. 1979. Isolation and characterization of the haemocytes of *Celliphora vicina* on density gradients of Ficoll. *J. Insect Physiol.* 25: 795-803.

Pelc, R. 1986. The haemocytes and their classification in the larvae and pupae of *Mamestra brassicae* (L.) 1758 (Lepidoptera: Noctuidae). *Can. J. Zool.* 64: 2503-2508.

Puttler, B. 1967. Interrelationship of *Hypera postica* (Coleoptera: Curculionidae) and *Bathyplectes curculionis* (Hymenoptera: Ichneumonidae) in the eastern United States with particular reference to encapsulation of the parasite eggs by the weevil larvae. *Ann. Ent. Soc. Am.* 60: 1031-1038.

Ratcliffe, N.A. and A.F. Rowley. 1979. Role of hemocytes in defence against biological agents. In: *Insect Hemocytes*, pp. 331-414 (A.P. Gupta, ed.). Cambridge University Press, Cambridge.

Renwarantz, L. 1986. Lectins in molluscs and arthropods: Their occurrence, origin and roles in immunity. In: *Immune Mechanisms in Invertebrate Vectors*, pp. 81-94 (A.M. Lackie, ed.). Oxford University Press, Oxford.

Rizki, M.T. 1957. Tumor formation in relation to metamorphosis in *Drosophila melanogaster*. *J. Morph.* 100: 459-472.

Rizki, M.T. 1960. Melanotic tumor formation in *Drosophila*. *J. Morph.* 106: 147-157.

Rizki, M.T. 1962. Experimental analysis of haemocyte morphology in insects. *Amer. Zool.* 2: 247-256.

Rizki, M.T. and R.M. Rizki. 1959. Functional significance of the crystal cells in the larvae of *Drosophila melanogaster*. *J. Biophys. Biochem. Cytol.* 5: 235-240.

Rowley, A.F. and N.A. Ratcliffe. 1981. Insects. In: *Invertebrate Blood Cells*, vol. 2, pp. 471-490 (N.A. Ratcliffe and A.F. Rowley, eds.). Academic Press, New York.

Rowley, A.F., N.A. Ratcliffe, C.M. Leonard, E.H. Richards and L. Renwarantz. 1986. Humoral recognition factors in insects, with particular reference to agglutinins and the pro-phenoloxidase system. In: *Hemocytic and Humoral Immunity in Arthropods*, pp. 381-406 (A.P. Gupta, ed.). John Wiley and Sons, New York.

Ryerse, J.S. 1986. The control of Malpighian tubule developmental physiology by 20-hydroxyecdysone and juvenile hormone. *J. Insect. Physiol.* 26: 449-457.

Salt, G. 1963. The defense reactions of insects to metazoan parasites. *Parasitology* 53: 527-612.

Schneider, F. 1950. Die Abwehrreaktion des Insektenblutes und ithre Beeinflussung durch die Parasiten. *Vischr. Naturf. Ges. Zurich* 95: 22-24.

Smith, S.L. 1985. Regulation of ecdysteroid titer: synthesis. In: *Comprehensive Insect Physiology, Biochemistry and Pharmacology*, vol. 7, pp. 295-341 (D.A. Kerkut and L.I. Gilbert, eds.). Pergamon Press, New York.

Söderhäll, K. and R.A. Jaxon. 1982. Effect of quinones and melanin on mycelial growth of *Aphanomyces* spp. and extracellular protease of *Aphanomyces astaci*, a parasite on crayfish. *J. Invertebr. Pathol.* 39: 105-109.

Söderhäll, K. and V.J. Smith. 1986. Pro-phenoloxidase-activating cascade as a recognition and defence system in arthropods. In: *Hemocytic and Humoral Immunity in Arthropods*, pp. 251-285 (A.P. Gupta, ed.). John Wiley and Sons, New York.

Stebbins, M.R. and K.D. Hapner. 1986. Isolation characterization and inhibition of arthropod agglutinins. In: *Hemocytic and Humoral Immunity in Arthropods*, pp. 99-146 (A.P. Gupta, ed.). John Wiley and Sons, New York.

Steele, C.G.H. and K.G. Davey. 1985. Integration in the insect endocrine system. In: *Comprehensive Insect Physiology, Biochemistry and Pharmacology*, vol. 8, pp. 1-35 (G.A. Kerkut and L.I. Gilbert, eds.). Pergamon Press, New York.

Steele, J.E. 1985. Control of metabolic processes. In: *Comprehensive Insect Physiology, Biochemistry and Pharmacology*, vol. 8. pp. 99-146, (G.A. Kerkut and Gilbert, L.I. eds.). Pergamon Press, New York.

Stoffolano, J.G. and F.A. Streams. 1971. Host reactions of *Musca domestica*, *Orthellis caesarion*, and *Ravinia l'herminieri* to the nematode *Hetrotylenchus autumnalis*. *Parasitology* 63: 195-211.

Van den Bosch, R. 1964. Encapsulation of the eggs of *Bathyplectes curculionis* (Thomson) (Hymenoptera: Ichneumonidae) in larvae of *Hypera brunneipennis* (Boheman) and *Hypera postica* (Gyllenhall) (Coleoptera: Curculionidae). *J. Insect Pathol.* 6: 343-367.

Van Driesche, R.G., A. Bellotti, C.J. Herrera and J.A. Castillo. 1986. Encapsulation rates of two encyrtid parasitoids by two *Phenacoccus* spp. of cassava mealybugs in Colombia. *Entomol. Exp. Appl.* 42: 79-82.

Vey, A. 1971. Richerches sur la reaction hemocytaire anticryplogamique de type granulome chez les insects. Doctoral Dissertation, Univ. Toulouse, France.

Wago, H. and Y. Ishikawa, 1979. Hemocytic reactions to foreign cells in the silkworm, *Bombyx mori* during post-embryonic development. *Appl. Ent. Zool.* 14: 36-43.

Walker, I. 1959. Die Abwehrreaktion des Wirtes *Drosophila melanogaster* gegen die zoophage Cynipide *Pseudeucoila bochei* Weld. *Rev. Suisse Zool.* 66: 569-632.

Walters, D.R. 1970. Hemocytes of saturniid silkworms: their behavior *in vivo* and *in vitro* in response to diapause, development, and injury. *J. Exp. Zool.* 134: 441-450.

Weaver, R.J. 1981. Juvenile hormone regulation of left colleterial gland function in the cockroach, *Periplaneta americana*: oothecin synthesis. In: *Juvenile Hormone Biochemistry*, pp. 271-289 (G.E. Pratt and G.T. Brooks, eds.). Elsevier/North-Holland Biomedical Press, Amsterdam.

Webley, D.P. 1951. Blood cell counts in the African migratory locust (*Locusta migratoria migratorioides* Rieche and Fairmaire). *Proc. Roy. Entomol. Soc.* (Lond). A26: 25-37.

Wheeler, R.E. 1963. Studies on the total hemocyte count and hemolymph volume in *Periplaneta americana* (L.) with special reference to the last molting cycle. *J. Insect Physiol.* 91: 223-235.

Wigglesworth, V.B. 1955a. The role of hemocytes in the growth and molting of an insect, *Rhodnius prolixus*. *O.J. Microsic. Sci.* 97: 89-98.

Wigglesworth, V.B. 1955b. The endocrine chain in an insect. *Nature* 175: 338.

Wigglesworth, V.B. 1964. The hormonal regulation of growth and reproduction in insects. *Adv. Insect Physiol.* 2: 248-335.

Wigglesworth, V.B. 1979. Hemocytes and growth in insects. In: *Insect Hemocytes*, pp. 303-318 (A.P. Gupta, ed.). Cambridge University Press, New York.

Yeager, J.F. 1945. The blood picture of Southern armyworm (Prodenia eridania). *J. Agric. Res.* 71: 1-40.

Zachary, D and J.A. Hoffmann. 1980. Endocrine control of the metamorphosis of the larval muscles in *Calliphora erythrocephala* (Diptera); *in vitro* studies of the role of ecdysteroids. *Devel. Biol.* 80: 235-247.

CHAPTER 9

Prophenoloxidase Activating System and Its Role in Cellular Communication

Kenneth Söderhäll and Anna Aspán

Introduction

Invertebrates are able to protect themselves from the harmful effects of invading parasites which may gain entry into the body cavity. In arthropods the cuticle forms a mechanical and chemical barrier to invading parasites (Söderhäll et al., 1988; Ratcliffe et al., 1985). If the parasite succeeds in reaching the haemocoel, the blood cells or haemocytes try to remove and destroy the foreign particle. The haemocytes can phagocytose smaller foreign entities, such as bacteria or fungal spores, but larger parasites, for example fungal hyphae, are encapsulated by several haemocytes and then removed from circulation (see Ratcliffe et al., 1985; Söderhäll and Smith, 1986a and b). In recent years it has been possible to demonstrate that the various haemocyte types in crustaceans are able to communicate between themselves by means of specific molecules and that the so-called prophenoloxidase activating system appears to play an important role in these processes. This review describes the role of the prophenoloxidase activating system (pro-PO system) in immune recognition and cellular communication.

Elicitors of the Prophenoloxidase Activating System

The last or terminal component of the pro-PO system is phenoloxidase, an enzyme which is responsible for the melanization reaction often observed as a response to parasites in the body cavity of arthropods. In 1977, Unestam and Söderhäll showed for the first time that phenoloxidase activity could be strongly enhanced by β-1,3-glucans from fungal cell walls and, further, demonstrated that the minimal degree of polymerization of a β-1,3-glucan to function as an elicitor of pro-PO system activation was 5 (Söderhäll and Unestam, 1979). Later, it was confirmed that β-1,3-glucans, in addition to other microbial polysaccharides, such as lipopolysaccharides and peptidoglucans, could activate the pro-PO system in several species of arthropods (see Söderhäll, 1982; Ashida et al., 1982; Ratcliffe et al., 1985). Thus it appears that the pro-PO system itself can react to foreign molecules and be converted into its active form by these microbial products.

Biochemistry of the Prophenoloxidase Activating System

Although a substantial number of papers have been published on the pro-PO system and its activation (Ratcliffe et al., 1985; Söderhäll and Smith, 1986a and b), still the details of pro-PO activation in arthropods are largely not known. However, in recent years some progress has been made in isolating and purifying proteins which are components of, or are associated with, the pro-PO system. Table 1 lists the prophenoloxidases which have been isolated and purified so far. Some may have been isolated in a partially active form since, as previously demonstrated in both the insect *Bombyx mori* (Ashida and Dohke, 1980) and a freshwater crayfish (Söderhäll et al., 1979), an active phenoloxidase is hydrophobic and tends to aggregate easily. The following discussion of the pro-PO system and its activation is limited to work done only with purified components isolated and identified in one insect and one crustacean.

Table 1: Prophenoloxidases purified from arthropods

	Molecular mass		Ip	References
	Prophenoloxidase	Phenoloxidase		
Calliphora erythrocephala	15.4 S[a]	N.D.		Karlson et al., 1964
	9.4 S[a]; 115 kDa[b]	N.D.		Munn and Bufton, 1973
	87 kDa[c]	N.D.		Naqvi and Karlson, 1979
Tenebrio molitor	7.3 S[a]	N.D.		Heyneman, 1965
Bombyx mori	37 kDa[a]; 80 kDa[c,d]		4.98	Ashida, 1971
		70 kDa[c]		Ashida and Yoshida, 1988
Manduca sexta	100-150 kDa[b];	N.D.		Aso et al., 1985
	71 kDa & 77 kDa[d]			
Musca domestica	178 kDa[b]	310-340 kDa[b]		Tsukamoto et al., 1986
Hyalophora cecropia	76 kDa[c,d]	N.D.		Andersson et al., 1989
Blaberus craniifer	76 kDa	ND		Durrant et. al., 1992
Pacifastacus leniusculus	300 kDa[b]; 76kDa[c,d]	60 kDa & 62 kDa[c]	5.4	Aspán and Söderhäll, 1991

Molecular masses determined by:
  a) sucrose density gradient centrifugation
  b) gel filtration chromatography
  c) SDS-polyacrylamide gel electrophoresis under non-reducing conditions
  d) SDS-polyacrylamide gel electrophoresis under reducing conditions
  N.D.—not determined

PRO-PO SYSTEM OF *BOMBYX MORI*

Ashida and colleagues have to date isolated and purified two components of the pro-PO system in the insect *B. mori*, namely, prophenoloxidase (Ashida, 1971) and a β-1,3-glucan binding protein (Ochiai and Ashida, 1988). The *B. mori* prophenoloxidase had a molecular mass of 80 kDa and could be turned into its active form by commercial proteinases. A serine proteinase isolated from the cuticle of *B. mori* (Dohke 1973a and b) could also activate pro-PO and, upon proteolytic activation of the insect pro-PO, a 5 kDa peptide was released (Ashida, 1974). However, it was apparent that the serine proteinase in the cuticle of this insect was not the native and natural activator and, recently, Ashida and colleagues found that the serine proteinase activity in the plasma is associated with pro-PO activation (Yoshida and Ashida, 1986; Ashida and Yoshida, 1988). In its active form the phenoloxidase of *B. mori* had a mass of about 70 kDa.

PRO-PO SYSTEM IN A CRUSTACEAN

The pro-PO we purified from the blood cells of a freshwater crayfish had a molecular mass of 76 kDa. It consisted of one polypeptide chain (Aspán and Söderhäll, 1991) in contrast to the pro-PO of *B. mori*, which is a dimer (Ashida, 1971). The enzyme responsible for conversion of the crayfish pro-PO to its active form, phenoloxidase, was a serine proteinase (ppA) with a molecular mass of 36 kDa (Fig. 1), but after reduction a 28 kDa fragment, containing the active site, and several smaller peptides were produced (Fig. 1) (Aspán *et al.*, 1990b). When the ppA and pro-PO were incubated together, the ppA turned the pro-PO into active phenoloxidase and, as a result, the molecular mass changed from 76 kDa to two peptides with masses of 60 and 62 kDa, and both peptides exhibited phenoloxidase activity (Fig. 2) (Aspán and Söderhäll, 1991). A commercial trypsin can also induce activation of pro-PO but only a 60 kDa phenoloxidase is produced (Fig. 2). Thus it is clear that, at least in this crustacean, prophenoloxidase can be activated by an endogenous proteinase and this activation appears to be a result of proteolytic cleavage of the pro-PO. However, in the blood cells of crayfish four serine proteinases or ^{3}H-DFP binding proteins are present, with masses of 36, 38, 50 and 67 kDa (Aspán and Söderhäll, 1991). Thus it seemed possible that some of the other serine proteinases might also be able to activate pro-PO. Experiments were therefore conducted, which provided conclusive evidence that the 36 kDa serine proteinase was indeed the prophenoloxidase-activating enzyme, (ppA). A crude source of pro-PO components, i.e., a blood cell lysate, was first separated on SDS-PAGE and then transferred to nitrocellulose (NC) filters. An inactive pro-PO-system was then added to the NC filter and, as shown in Fig. 3, the prophenoloxidase became bound to two proteins with masses of 36 and 38 kDa (Aspán and Söderhäll, 1991). The 38 kDa serine proteinase was present in a much lower quantity than the 36 kDa and although both may

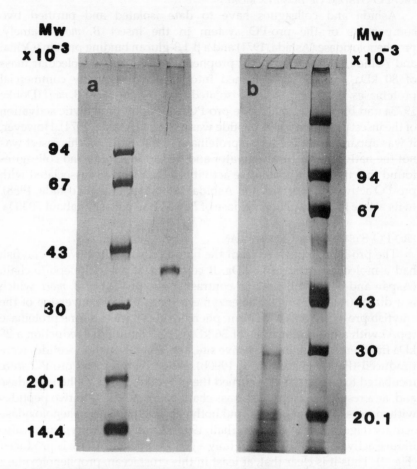

Fig. 1: SDS-PAGE of the purified 36 kDa serine proteinase, the prophenoloxidase activating enzyme. a—non-reduced sample (15 µg); b—reduced sample (15 µg). The active site of this ppA is located on the 28 kDa fragment. The gel was stained with Coomassie brilliant blue. *Source*: Aspán *et al.* (1990b).

convert pro-PO to active PO, it appears likely that the 36 kDa plays a more important role as a ppA in the haemolymph of the freshwater crayfish.

The 36 kDa ppA exhibited no enzymatic activity in the blood cells of crayfish (Söderhäll, 1983) until elicitors of the pro-PO system activation were added. This could indicate that the ppA is present in an inactive form, maybe as a zymogen. Further evidence to suggest that the 36 and 38 kDa serine proteinase are present in enzymatically inactive forms in crayfish blood cells is that in a whole haemocyte lysate incubated with ^{3}H-DFP, only the 50 and 67 kDa proteinases were labelled. On the other hand, when laminarin, a β-1,3-glucan, was added to the haemocyte lysate, four proteinases incorporated ^{3}H-DFP, namely 36, 38, 50 and 67 kDa (Aspán and Söderhäll, 1991). Together with the fact that in a non-activated

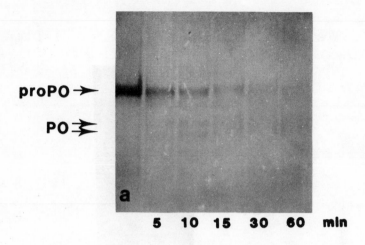

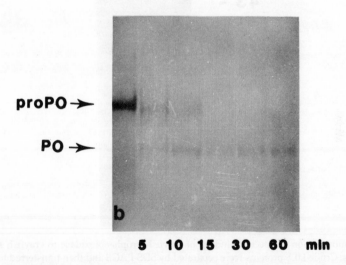

Fig. 2: Proteolytic cleavage of a homogeneous prophenoloxidase from crayfish blood cells. a—crayfish pro-PO incubated with a pure ppA from crayfish; b—crayfish pro-PO incubated with trypsin. The gel was stained with Coomassie brilliant blue.
Source: Aspán and Söderhäll (1991).

haemocyte lysate no proteinase activity towards Bz-lle-Glu-(γ-O-Piperidyl)-Gly-Arg-para-nitroaniline (a good substrate for the 36 and 38 kDa serine proteinases) is detected, this strongly suggests that the 36 and 38 kDa serine proteinases are present in an inactive form within the haemocytes. Whether the ppA is a zymogen awaits further study, however.

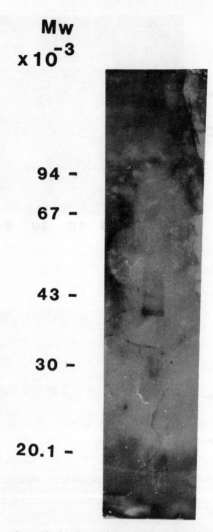

Fig. 3: Demonstration of specific binding of crayfish prophenoloxidase to crayfish serine proteinases. Crude HLS proteins were separated by SDS-PAGE and then transferred to NC-filters. The filters were then incubated with a fresh haemocyte lysate containing prophenoloxidase. Phenoloxidase activity of the bound prophenoloxidase was demonstrated using L-DOPA (L-dihydroxyphenylalanine) as the substrate.
Source: Aspán and Söderhäll (1991).

In crustaceans, the pro-PO system can also be triggered into its active form without the presence of microbial polysaccharides if the Ca^{2+} concentration is lowered (Söderhäll, 1981; Söderhäll and Häll, 1984) and a possible function for this mechanism is that the pro-PO system may also participate in wounding responses (Söderhäll, 1981, Durliat, 1985).

INITIAL STAGES IN PRO-PO SYSTEM ACTIVATION IN ARTHROPODS

Since the pro-PO system in both insects and crustaceans can be converted into its active form by microbial polysaccharides, such as β-1,3-glucans, lipopolysaccharides or peptidoglucans, it is not surprising that proteins with the capacity to bind these elicitors have been identified. A β-1,3-glucan binding protein has been isolated and purified from two insect species, namely *Bombyx mori* (Ochiai and Ashida, 1988) and *Blaberus craniifer* (Söderhäll *et al.*, 1988), and also from one crustacean, *Pacifastacus leniusculus* (Duvic and Söderhäll, 1990). After binding to a β-1,3-glucan, these proteins enhanced the activity of both ppA and phenoloxidase in these animals. The exact mechanism by which this activation was achieved is still not known. Proteins with the capacity to bind other elicitors of the pro-PO system activation await isolation and characterization.

RELEASE OF PRO-PO SYSTEM FROM BLOOD CELLS IN CRUSTACEANS

In arthropods the pro-PO system is located either in the haemocytes, as in crustaceans and in some insects, or in the plasma, as claimed for other insect species (see Söderhäll and Smith, 1986a and b; Ratcliffe *et al.*, 1985). The exact localization of the pro-PO system in various insect species awaits more detailed studies, but it is clear that in crustaceans the whole system is localized within the haemocytes and, more specifically, in the vesicles of the semi-granular and granular cells (Johansson and Söderhäll, 1989). This means that in crustaceans this system has to be released from the cells before it can become activated (Johansson and Söderhäll, 1985, 1989) and hence is subject to control and regulation.

Release of the pro-PO system can be achieved by elicitors of pro-PO system activation, such as lipopolysaccharides or β-1,3-glucans. These microbial polysaccharides induced exocytosis of the semi-granular cells in crayfish and thus caused a discharge of the pro-PO system (Johansson and Söderhäll, 1985) while the granular cells remained unaffected (Johansson and Söderhäll, 1985). Since the granular cell in crustaceans is the main repository for the pro-PO system (Smith and Söderhäll, 1983; Söderhäll and Smith, 1983) it is also necessary that these cells be able to discharge their contents of pro-PO components in order to participate in the host's defence. It was recently demonstrated that two proteins could elicit this release from the granular cells, namely a 76 kDa factor, which is associated with the pro-PO system (Johansson and Söderhäll, 1988, 1989), and also the β-1,3-glucan binding protein if previously reacted with a β-1,3-glucan (Barracco *et al.*, 1991). This means that two proteins of the pro-PO system have been shown to be directly involved in the communication between cells in a crustacean. Less is known about the cell-to-cell communication in insects.

CONTROL OF PRO-PO SYSTEM ACTIVATION IN ARTHROPODS

Once outside the cells, activation of the pro-PO system can be controlled by proteinase inhibitors since in the plasma of crayfish two high molecular weight proteinase inhibitors can prevent activity of ppA and thereby block pro-PO activation. These inhibitors have been purified and characterized: one is an α_2-macroglobulin (Hall *et al.*, 1989) and the other a 155 kDa trypsin inhibitor (Hergenhahn *et al.*, 1987). The 155 kDa proteinase inhibitor is more efficient in inhibiting ppA (Aspán *et al.*, 1990a) (Table 2).

Proteinase inhibitors are also present in insect haemolymph (Saul and Sugumaran, 1986) and may be involved in regulating the pro-PO system activation. So far no such inhibitor has been isolated and characterized.

Table 3 summarizes our present knowledge of isolated components or associated factors of the pro-PO system as they occur in one crustacean, the freshwater crayfish, *P. leniusculus*, and Fig. 4 shows a model for activation of the pro-PO system as it occurs in this arthropod.

Table 2: Proteinase inhibitors and their function in crustacean haemolymph

	Mass kDa	Biological function	References
Subtilisin inhibitor	23	Inhibits microbial proteinases	Häll and Säderhäll, 1982
α-macroglobulin	2 × 190	Inhibits ppA in a molar ratio of 1:25	Hergenhahn *et al.*, 1988 Aspán *et al.*, 1990a
Trypsin inhibitor	155	Inhibits ppA in a molar ratio of 1:2.5	Hergenhahn *et al.*, 1987 Aspán *et al.*, 1990a

Table 3: Proteins associated with the prophenoloxidase activating system of crayfish

Protein	Molecular mass	Localization	References
Prophenoloxidase	76 kDa	Haemocytes	Aspán and Söderhäll, 1991
Prophenoloxidase-activating enzyme	36 kDa	Haemocytes	Aspán *et al.*, 1990b
Gelatinase	50 kDa	Haemocytes	Aspán and Söderhäll, 1991
67 kDa proteinase	67 kDa	Haemocytes	Aspán and Söderhäll, 1991
Subtilisin inhibitor	23 kDa	Haemocytes	Häll and Söderhäll, 1982
Trypsin inhibitor	155 kDa	Plasma	Hergenhahn *et al.*, 1987
α₂-macroglobulin	2 × 190 kDa	Plasma	Hergenhahn *et al.*, 1988 Aspán *et al.*, 1990a
Cell adhesion, degranulation and encapsulation promoting factor	76 kDa	Haemocytes	Johansson and Söderhäll, 1988 Johansson and Söderhäll, 1989 Kobayashi *et al.*, 1990
β-1,3-glucan binding protein	100 kDa	Plasma	Duvic and Söderhäll, 1990
Interleukin 1-like protein	25 kDa	Haemocytes	Aspán and Söderhäll, unpublished

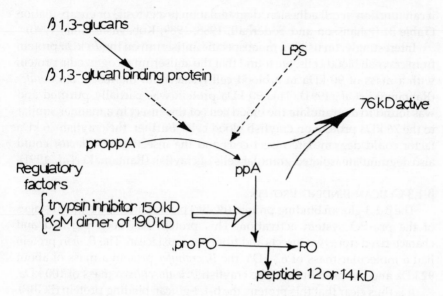

Fig. 4: Model for pro-PO system activation in arthropods, especially freshwater crayfish. *Source*: Söderhäll (1992).

Biological Functions of Different Factors of the pro-PO system

ANTIFUNGAL ACTIVITY OF PRODUCTS OF PHENOLOXIDASE ACTIVITY

The terminal component of the pro-PO cascade is prophenoloxidase (Aspán and Söderhäll, 1991) which, in its active form, is a redox enzyme capable of oxidizing phenols to quinones which later spontaneously and without enzyme catalysis form melanin as an end product. Söderhäll and Ajaxon (1982) showed that melanin and some intermediates in the melanin biosynthesis were inhibitory to both growth of and proteinase activity of some parasitic fungi. Later, other investigators also showed that these products could affect fungal growth (St. Leger *et al.*, 1988; Rowley *et al.*, 1990).

76 KDA PROTEIN IN CRUSTACEANS

As mentioned above, a 76 kDa protein can cause exocytosis of the pro-PO system from both the semi-granular and granular blood cells of crayfish (Johansson and Söderhäll, 1989). This protein was purified from the blood cells (Johansson and Söderhäll, 1988). The 76 kDa factor is biologically inactive in the blood cells and its activity is generated concomitant with the activation of the pro-PO system (Johansson and Söderhäll, 1988). As yet we do not know how this is achieved but it would seem that a conformational change in the 76 kDa factor is required to render it biologically active. The 76 kDa factor has three biological activities since it can function in cell adhesion, degranulation (exocytosis) or encapsulation (Table 3) (Johansson and Söderhäll, 1988, 1989; Kobayashi *et al.*, 1990)

Interestingly, on using a monospecific antiserum on the 76 kDa protein from crayfish blood cells, we found that the antiserum recognized a protein with a mass of 90 kDa in a blood cell lysate from the insect *B. craniifer* (Rantamäki *et al.*, 1991). This 90 kDa protein was partially purified and was found to degranulate the blood cells of this insect in a manner similar to the 76 kDa protein on crayfish blood cells. Further, the crayfish 76 kDa factor could degranulate insect cells and the insect 90 kDa factor could also degranulate isolated granular cells of crayfish (Rantamäki *et al.*, 1991).

β-1,3-GLUCAN BINDING PROTEIN

The β-1,3-glucan binding protein (BGBP) participates in the initial stages of the pro-PO system activation. This protein has been purified and characterized from two insects and from one crustacean. The *B. mori* protein had a molecular mass of 62 kDa, the *B. craniifer* protein a mass of about 92 kDa and the protein from the crayfish *P. leniusculus* a mass of 100 kDa.

It is thus clear that this protein, the β-1,3-glucan binding protein (BGBP), has a biological function, since by binding to fungal β-1,3-glucans in some unknown way, it induces activation of the pro-PO system or, more correctly, enhances the activity of ppA and phenoloxidase. However, in crayfish plasma we found that the concentration of BGBP far exceeded that required solely for involvement in pro-PO activation. It was recently demonstrated that the BGBP in freshwater crayfish, in addition to functioning in the pro-PO system activation, is also important in the communication between blood cells, since a BGBP previously treated with a β-1,3-glucan induced spreading and partial degranulation of crayfish granular blood cells. Untreated BGBP, on the other hand, had no such activity (Fig. 5) (Barracco *et al.*, 1991). It has also been shown, again in crayfish, that a protein associated with the pro-PO system is involved in cellular communication, as was previously shown with the 76 kDa factor.

Cell-to-Cell Communication in Arthropods

Perhaps the best way to rapidly advance our understanding of how different blood cells communicate amongst themselves during an infection, is to use separated and isolated blood cells. In 1983, we developed a method to isolate and handle crustacean blood cells *in vitro* (Söderhäll and Smith, 1983; Smith and Söderhäll, 1983). The method is quite simple and the important factor in maintaining the integrity of the blood cells is to bleed the haemolymph into an anticoagulant with a low pH (around 4.6) and containing EDTA. After collecting the haemolymph in this anticoagulant, the blood cells are separated on preformed Percoll gradients, harvested and then used for *in vitro* studies. The same technique was later applied

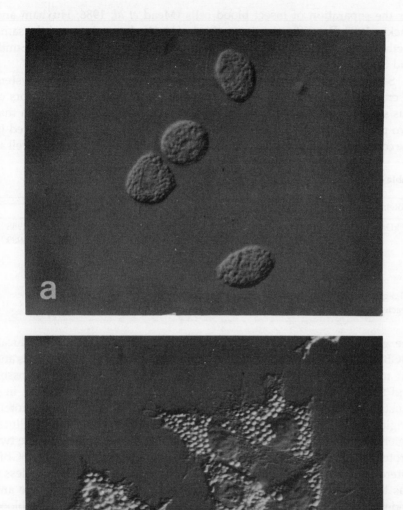

Fig. 5: Effect of the β-1,3-glucan binding protein on monolayers of crayfish granular cells. a—unaffected cells treated with the β-1,3-glucan binding protein; b—spread and partially degranulated cells incubated with the β-1,3-glucan binding protein previously treated with laminarin (a β-1,3-glucan) (magnification 400 X).
Source: Barracco *et al.* (1991).

for the separation of insect blood cells (Mead *et al*, 1986; Huxham and Lackie, 1988; Ratcliffe, chapter in this book). We have also used the same method to isolate blood cells from a wide range of invertebrate phyla (Smith and Söderhäll, 1991).

Separated blood cells and purified components from the pro-PO system of crayfish were used to study the role, if any, of the various factors of this system in cell-to-cell communication. To date, we have shown that two proteins associated with the pro-PO system are directly involved in the communication between cells (Table 4). Both the 76 kDa factor as well as

Table 4: Proteins involved in cell-to-cell communication in freshwater crayfish

Protein	Biological function	References
76 kDa	Cell adhesion factor	Johansson and Söderhäll, 1988
	Degranulation factor	Johansson and Söderhäll, 1989
	Encapsulating promoting factor	Kobayashi *et al.*, 1990
β-1,3-glucan binding protein (reacted with β-1,3-glucan)	Degranulation factor Cell-spreading factor	Barracco *et al.*, 1991

the glucan binding protein, if previously reacted with a β-1,3-glucan (BGBP-L), can cause spreading and degranulation of the granular cells and the mechanism by which this occurs is regulated exocytosis (Johansson and Söderhäll, 1988, 1989; Barracco *et al.*, 1991). This means that in a crustacean one protein present in the plasma, the BGBP, and one protein originally present in the blood cells are capable of participating in cellular defence. Presently, we are identifying membrane receptors for these two proteins and, in fact, have already isolated such a receptor for the BGBP. Interestingly, the BGBP does not bind to its membrane receptor unless it has been previously reacted or treated with a β-1,3-glucan (Duvic and Söderhäll, 1992), which accords with the binding of BGBP to intact blood cells (Barracco *et al.*, 1991).

Although we have begun to unravel some of the complex reactions occuring during cellular defence in crustaceans, much still remains unknown and it is essential to begin studies on the molecular processess involved in cellular defences, as well as which secondary messenger systems are induced to operate during, for example, exocytosis in crayfish granular cells induced by the BGBP. Such studies are presently under way in our laboratory.

Unfortunately, much less is known about the role, if any, of the pro-PO system in cellular communication processes in insects. Most work so far has indicated that the insect pro-PO system may have a function similar to that of crustaceans. We have recently shown that a protein with a mass

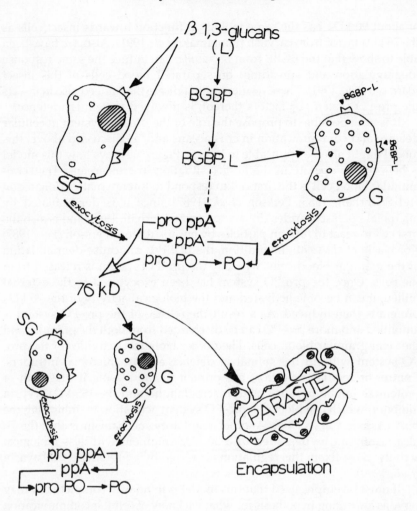

Fig. 6: *Cell to cell communication in crustaceans.* The semi-granular cell (SG-cell) is the first cell to recognize foreign particles and molecules such as fungal β-1,3-glucans. The β-1,3-glucans bind to receptors on the SG cells and as a result the pro-PO system contained in the vesicles of these cells is released by exocytosis to the plasma. Outside the cells the pro-PO system is turned into its active form by complexes of the β-1,3-glucan binding protein (BGBP) and β-1,3-glucan. As a result of the pro-PO system activation, the 76 kDa factor obtains its biological activity and the 76 kDa can then release more pro-PO system from both the SG or the granular (G) cells. In addition, the BGBP, if reacted with a β-1,3-glucan, can also trigger release of more pro-PO system from the G cells. In this way the release of the pro-PO system is amplified and, after being activated outside the cells can participate in cellular defence by providing opsonins for encapsulating cells. Outside the cells (i.e., in the plasma) proteinase inhibitors are present which can inhibit the activity of ppA and thus block the pro-PO system activation in places where it is inappropriate. It has to be emphasized though that much is still not known about the molecular details of this cell-to-cell communication. *Source*: Söderhäll (1992).

of about 90 kDa has the same biological function towards insect cells as the 76 kDa factor from caryfish (Rantamäki *et al.*, 1991). Also, we have been able to show that the BGBP from *B. craniifer* can induce the same response (degranulation and spreading) on separated blood cells of this insect (Barracco *et al.*, 1991). These results could indicate that indeed also in insects the pro-PO system, or factors thereof, is involved in cellular defence.

It is now possible to propose the role of the pro-PO system in cellular defence and communication in crustaceans and, as discussed above, this model might also be applicable to insects. Figure 6 shows how this model is believed to operate in the *in vivo* situation in crustaceans. The semi-granular blood cell is the first cell to respond to foreign matter (Johansson and Söderhäll, 1985; Persson *et al.*, 1987) since it is degranulated by microbial polysaccharides (Johansson and Söderhäll, 1985) and is also the first cell to react to foreign particles by encapsulation (Persson *et al.*, 1987; Kobayashi *et al.*, 1990). In addition, the BGBP-L can cause degranulation of the granular blood cells, whereby the pro-PO system is released from the cells. Once the pro-PO system has been exocytosed to the external milieu, it can become activated and the biological activity of the 76 kDa protein is then induced. As a result the release of the pro-PO system is amplified and more pro-PO can be discharged from both the granular and the semi-granular blood cells. Then other bioligical activities of the pro-PO system will aid the animal in defence against microbial intruders. Control or regulatory factors are present in the plasma in the form of proteinase inhibitors, such as α_2-macroglobulin and the 155 kDa trypsin inhibitor, which can block the pro-PO system activation by inhibiting the ppA (Aspán *et al.*, 1990a). Another regulatory mechanism is that the 76 kDa factor can be degraded to a 30 kDa fragment with less biological activity; most likely this regulation is exerted by a proteinase (Kobayashi *et al.*, 1990).

It must be emphasized that this model is in no way complete but may serve as a working hypothesis for what we know of cellular communication in crustaceans. Hopefully we will soon be able to provide this model with more detailed information on the molecular processes involved. As already said, it will be interesting to see whether cell-to-cell communication occurs in a similar manner in insects.

Conclusions

In recent years we have been successful in isolating and purifying six proteins which are components or associated factors of the pro-PO system in crustaceans. The role of these factors in cellular defence and communication has been and still is under study. Thus we are only in the early stages of knowing how different blood cells communicate amongst themselves during an infection. It is our hope to unravel in the not too distant future, more details about the molecular and biochemical processes involved in cell-to-cell communication as they occur in crustaceans and insects.

ACKNOWLEDGEMENT

Most of the original research reported in this review was financed by the Swedish Natural Science Research Council.

REFERENCES

Andersson K., S.-C. Sun, H.G. Boman and H. Steiner. 1989. Purification of Cecorpia prophenoloxidase and four proteins involved in its activation. *Insect Biochem.* 19: 629-638.

Ashida, M. 1971. Purification and characterization of prephenoloxidase from hemolymph of the silkworm, *Bombyx mori. Arch. Biochem. Biophys.* 144: 749-762.

Ashida, M. 1974. Activation of prephenoloxidase. III. Release of a peptide from prephenoloxidase by the activating enzyme. *Biochem. Biophys. Res. Commun.* 57: 1089-1095.

Ashida, M. and H. Yoshida. 1988. Limited proteolysis of prophenoloxidase during activation by microbial products in insect plasma and effect of phenoloxidase on electrophoretic mobilities of plasma proteins. *Insect Biochem.* 18: 11-19.

Ashida, M. and K. Dohke. 1980. Activation of prophenoloxidase by the activating enzyme of the silkworm, *Bombyx mori. Insect Biochem.* 10: 37-47.

Ashida, M., R. Iwama, H. Iwahana and H. Yoshida. 1982. Control and function of the prophenoloxidase activating system. In: *Proceedings of the 3rd International Colloquium on Invertebrate Pathology*, pp. 81-86 (C.C. Payne and H.D. Burges, eds.). University of Sussex, Brighton.

Aso, Y., K.J. Kramer, T.L. Hopkins and G.L. Lookhart. 1985. Characterization of haemolymph protyrosinase and a cuticular activator from *Manduca sexta* (L.) *Insect Biochem.* 15: 9-17.

Aspán, A., J. Sturtevant, V.J. Smith and K. Söderhäll. 1990b. Purification and characterization of a prophenoloxidase activating enzyme from crayfish blood cells. *Insect Biochem.* 20: 709-718.

Aspán, A. and K. Söderhäll. 1991. Purification of prophenoloxidase from crayfish blood cells and its activation by an endogenous serine proteinase. *Insect Biochem.* 21: 363-373.

Aspán, A., M. Hall and K. Söderhäll. 1990a. The effect of endogenous proteinase inhibitors on the prophenoloxidase activating enzyme, a serine proteinase from crayfish haemocytes. *Insect Biochem.* 20: 485-492.

Barracco, M.A., B. Duvic and K. Söderhäll. 1991. The β-1,3-blucan binding protein, from plasma of *Pacifastacus leniusculus*, induces spreading and partial degranulation of crayfish granular cell. *Cell Tiss. Res.* 266: 491-497.

Dohke, K. 1973a. Studies on prephenoloxidase-activating enzyme from cuticle of the silkworm *Bombyx mori*. I. Activation reaction by the enzyme. *Arch. Biochem. Biophys.* 157: 203-209.

Dohke, K. 1973b. Studies on prephenoloxidase-activating enzyme from cuticle of the silkworm *Bombyx mori*. II. Purification and characterization of the enzyme. *Arch. Biochem. Biophys.* 157: 210-221.

Durliat, M. 1985. Clotting processes in crustacea decapoda. *Biol. Rev.* 60: 473-498.

Durrant, H.J., N.A. Ratcliffe, C.R. Hipkin, A. Aspán and K. Söderhäll. 1992. Purification of the prophenoloxidase enzyme from haemocytes of the cockroach, *Blaberus discordales J. Biochem.* (in press).

Duvic, B. and K. Söderhäll. 1990. Purification and characterization of a β-1,3-glucan binding protein from plasma of the crayfish *Pacifastacus leniusculus. J. Biol. Chem.* 265: 9327-9332.

Duvic, B. and K. Söderhäll. 1992. Purification and partial characterization of a β-1,3-glucan-binding-protein membrane receptor from blood cells of the crayfish *Pacifastacus leniusculus. Eur. J. Biochem.* 207: 223-228.

Häll, L. and K. Söderhäll. 1982. Purification and properties of a protease inhibitor from crayfish hemolymph. *J. Invertebr. Path.* 39: 29-37.

Hall, M., K. Söderhäll. and L. Sottrup-Jensen. 1989. Amino acid sequence around the thiolester of α_2-macroglobulin from plasma of the crayfish *Pacifastacus leniusculus. FEBS Lett.* 254: 111-114.

Hergenhahn, H.G., A. Aspán and K. Söderhäll. 1987. Purification and characterization of a high-Mr proteinase inhibitor of pro-phenoloxidase activation from crayfish plasma. *Biochem. J.* 248: 223-228.

Hergenhahn, H.G., M. Hall and K. Söderhäll. 1988. Purification and characterization of an α_2-macroglobulin-like proteinase inhibitor from plasma of the crayfish *Pacifastacus leniusculus. Biochem. J.* 255: 801-806.

Heyneman, R.A. 1965. Final purification of a latent phenolase with mono- and diphenoloxidase activity from *Tenebrio molitor. Biochem. Biophys. Res. Comm.* 21: 162-169.

Huxham, I.M. and A.M. Lackie. 1988. Behaviour *in vitro* of separated haemocytes from the locust, *Schistocerca gregaria. Cell Tiss. Res.* 251: 677-684.

Johansson, M.W. and K. Söderhäll. 1985. Exocytosis of the prophenoloxidase activating system from crayfish haemocytes. *J. Comp. Physio. B* 156: 175-181.

Johansson, M.W. and K. Söderhäll. 1988. Isolation and purification of a cell adhesion factor from crayfish blood cells. *J. Cell Biol.* 106: 1795-1803.

Johansson, M.W. and K. Söderhäll. 1989. A cell adhesion factor from crayfish haemocytes has degranulating activity towards crayfish granular cells. *Insect Biochem.* 19: 183-190.

Karlson, P., D. Mergenhagen and C.E. Sekeris. 1964. Zum tyrosinstoffwechsel der insekten, XV. Weitere untersuchungen über das o-diphenoloxydase-system von *Calliphora erythrocephala. Hoppe-Seylers Z. Phisyol. Chemie.* 338: 42-50.

Kobayashi, M., M.W. Johansson and K. Söderhäll. 1990. The 76 kD cell adhesion factor from crayfish haemocyte promotes encapsulation *in vitro. Cell Tiss. Res.* 260: 13-18.

Mead, G.P., N.A. Ratcliffe and L.R. Renwrantz. 1986. The separation of insect haemocyte types on Percoll gradients: methodology and problems. *J. Insect Physiol.* 32: 167-177.

Munn, E.A. and S.F. Bufton. 1973. Purification and properties of a phenoloxidase from the blowfly *Calliphora erythrocephala. Eur. J. Biochem.* 35: 3-10.

Naqvi, S.N.H. and P. Karlson. 1979. Purification of prophenoloxidase in the haemolymph of *Calliphora vicina* (R.&D.). *Arch. Internat. Physiol. Biochem.* 87: 687-695.

Ochiai, M. and M. Ashida. 1988. Purification of a β-1,3-glucan recognition protein in the prophenoloxidase activating system from hemolymph of the silkworm, *Bombyx mori. J. Biol. Chem.* 263: 12056-12062.

Persson, M., A. Vey and K. Söderhäll. 1987. Encapsulation of foreign particles *in vitro* by separated blood cells from crayfish, *Astacus leptodactylus. Cell Tiss. Res.* 246: 409-415.

Rantamäki, J., H. Durrant, Z. Liang, N.A. Ratcliffe, B. Duvic and K. Söderhäll. 1991. Isolation of a 90 kD protein from haemocytes of *Blaberus craniifer*, which has similar functional and immunological properties to the 76 kD protein from crayfish haemocytes. *Insect Physiol* 37: 627-634.

Ratcliffe, N.A., A.F. Rowley, S.W. Fitzgerald and C.P. Rhodes. 1985. Invertebrate immunity: basic concepts and recent advances. *Int. Rev. Cytol.* 97: 183-350.

Rowley, A.F., J.L. Brookman and N.A. Ratcliffe. 1990. Possible involvement of the prophenoloxidase system of the locust, *Locusta migratoria*, in antimicrobial activity. *J. Invertebr. Path.* 56: 31-38.

Saul, S. and M. Sugumaran. 1986. Protease inhibitor controls prophenoloxidase activation in *Manduca sexta. FEBS Lett.* 208: 113-116.

Smith, V.J. and K. Söderhäll. 1983. Induction of degranulation and lysis of·haemocytes in the freshwater crayfish, *Astacus astacus*, by components of the prophenoloxidase activating system *in vitro. Cell Tiss. Res.* 233: 295-303.

Smith, V.J. and K. Söderhäll. 1991. A comparison of phenoloxidase activity in the blood of marine invertebrates. *Dev. Comp. Immunol.* 15: 251-261.

Söderhäll, K. 1981. Fungal cell wall β-1,3-glucans induce clotting and phenoloxidase attachment to foreign surfaces of crayfish hemocyte lysate. *Dev. Comp. Immun.* 5: 565-573.

Söderhäll, K. 1982. Prophenoloxidase activating system and melanization—a recognition mechanism of arthropods? A review. *Dev. Comp. Immun.* 6: 601-611.

Söderhäll, K. 1983. β-1,3-glucan enhancement of protease activity in crayfish hemocyte lysate. *Comp. Biochem Physiol. B* 74: 221-224.

Söderhäll, K. 1992. Biochemical and molecular aspects of cellular communication in arthropods. *Boll. Zoll. Ital.* 59: 141-151.

Söderhäll, K. and L. Häll. 1984. Lipopolysacharide-induced activation of prophenoloxidase activating system in crayfish haemocyte lysate. *Biochim. Biophys. Acta* 797: 99-104.

Söderhäll, K., L. Häll, T. Unestam and L. Nyhlén. 1979. Attachment of prophenoloxidase to fungal cell walls in arthropod immunity. *J. Invertebr. Pathol.* 34: 285-294.

Söderhäll, K. and R. Ajaxon. 1982. Effect of quinones and melanin on mycelial growth of *Aphanomyces* spp. and extracellular protease of *Aphanomyces astaci*, a parasite on crayfish. *J. Invertebr. Path.* 39: 105-109.

Söderhäll, K. and T. Unestam. 1979. Activation of serum prophenoloxidase in arthropod immunity. The specificity of cell wall glucan activation and activation by purified fungal glycoproteins of crayfish phenoloxidase. *Can. J. Microbiol.* 25: 406-414.

Söderhäll, K. and V.J. Smith. 1983. Separation of the haemocyte populations of *Carcinus maenas* and other marine decapods, and prophenoloxidase distribution. *Dev. Comp. Immunol.* 7: 229-239.

Söderhäll, K. and V.J. Smith. 1986a. The prophenoloxidase activating system: The biochemistry of its activation and role in arthropod cellular immunity, with special reference to crustaceans. In: *Immunity in Invertebrates*, pp. 208-223. (M. Brehélin, ed.). Springer-Verlag, Berlin.

Söderhäll, K. and V.J. Smith. 1986b. The prophenoloxidase activating cascade as a recognition and defence system in arthropods. In: *Humoral and Cellular Immunity in Arthropods*, pp. 251-285. (A.P. Gupta, ed.). John Wiley and Sons, New York.

Söderhäll, K., W. Rögener, I. Söderhäll, R.P. Newton and N.A. Ratcliffe. 1988. The properties and purification of a *Blaberus craniifer* plasma protein which enhances the activation of haemocyte prophenoloxidase by a β-1,3-glucan. *Insect Biochem.* 18: 323-330.

St. Leger, R.J., R.M. Cooper and A.K. Charnely. 1988. The effect of melanization of *Manduca sexta* cuticle on growth and infection by *Metarhizium anisopliae*. *J. Invertebr. Path.* 52: 459-470.

Tsukamoto, T., M. Ishiguro and M. Funatsu. 1986. Isolation of latent phenoloxidase from prepupae of the housefly, *Musca domestica*. *Insect Biochem.* 16: 573-581.

Unestam, T. and K. Söderhäll. 1977. Soluble fragments from fungal cell walls elicit defence reactions in crayfish. *Nature* 267: 45-46.

Yoshida, H. and M. Ashida. 1986. Microbial activation of two serine enzymes and prophenoloxidase in the plasma fraction of hemolymph of the silkworm, *Bombyx mori*. *Insect Biochem.* 16: 539-545.

CHAPTER 10

Some Biochemical Aspects of Eumelanin Formation in Insect Immunity

A. J. Nappi and M. Sugumaran

1. Introduction

Insects employ an arsenal of defences to successfully combat a diversity of nonself components. Their immune responses include phagocytosis, nodule formation, agglutination, secretion of antibacterial proteins, and cellular encapsulation (Ashida and Yamazaki, 1990; Boman *et al.*, 1991; Christensen and Nappi, 1988; Coombe *et al.*, 1984; Götz, 1986; Götz and Boman, 1985; Gupta, 1988; Lackie, 1988; Ratcliffe *et al.*, 1985; Salt, 1970; Schmidt and Theopold, 1991; Söderhäll, 1982; Sugumaran, 1990). Presently, very little is understood about the initial events of immune recognition, or how the various cellular and biochemical effector responses potentiated by the recognition process kill foreign organisms.

A common feature of encapsulation reactions made against foreign objects too large to be phagocytosed is the production of a brown to black pigment considered to be eumelanin, or a composite of eumelanin and sclerotin, although the physical and chemical properties of these capsules have never been adequately ascertained. Implicated in insect-host responses that sequester endoparasites in cellular, pigmented capsules are catecholamines and other biogenic amines derived from the metabolism of tyrosine (Nappi and Christensen, 1987; Nappi *et al.*, 1987, 1991, 1992a, b; Li *et al.*, 1989, 1992; Beckage *et al.*, 1990). Considering the potential damage to host tissues that can result from various quinone and hydroquinone intermediates generated during melanogenesis, especially in insects with their open circulatory system, elaborate homeostatic controls must operate to regulate the activities of the enzymes when needed for defence, and to integrate these functions with normal developmental processes requiring the same enzymes. Other well-known functions served by catecholamines include neurotransmission, cuticular sclerotization, and wound-healing responses. Thus, it would appear that the cellular melanotic encapsulation reactions in insects involve the collaborative interaction of enzymes that not only directly metabolize catecholamines and transient quinonoid compounds, but also degrade the storage forms of these precursors.

The purpose of this review is to examine some of the biochemical aspects of melanotic encapsulation occurring in immune reactive insects and to address the issue of putative killing molecules associated with this response.

2. Formation of Capsule Components

Endoparasites of insects typically are sequestered within cellular, pigmented capsules that historically have been referred to as "melanotic capsules". The cells participating in the encapsulation process are blood cells or hemocytes, but virtually nothing is known of the factors that stimulate cell proliferation in response to infections and to target foreign surfaces. The term "melanin" is not very descriptive, denoting a diverse group of natural and synthetic phenolic-quinonoid pigments but conveying no information as to the chemical nature of the pigment. In fact, black pigmentation in insects may represent widely different chemical entities even at successive developmental stages in the same species. Melanins may be classified as eumelanins, phaeomelanins and allomelanins (Nicolaus, 1968; Prota, 1988). This classification is based on known or presumed chemical precursors, differences in solubility, color, and type of degradation products. In addition to dopa, other metabolites of tyrosine that can give rise to melanin include dopamine, norepinephrine and epinephrine; even tryptophan metabolites can serve in this capacity (Kayser, 1985).

Eumelanins are typically brown or black, insoluble, heteropolymers comprising various o-diphenols and o-quinones synthesized from tyrosine and its hydroxylated derivative, dopa. These pigments contain high concentrations of oxidatively coupled indole monomers derived from 5, 6-dihydroxyindole (DHI) and 5,6-dihydroxyindole-2-carboxylic acid (DHICA). Since the polymerization of eumelanin appears to be essentially a random process, the final heteropolymeric product may comprise various other intermediates in the melanogenesis pathway, including dopa, dopaquinone, leucodopachrome, dopachrome, DHI and 5,6-indolequinone-2-carboxylic acid (IQCA) and 5,6-indolequinone (IQ) (Fig. 1). In addition, eumelanins are frequently found conjugated with proteins. A small amount of sulfur is usually present in eumelanin, indicating a possible conjugation to protein-bound cysteine and/or the presence of phaeomelanin.

Phaeomelanins are alkali-soluble, yellow to reddish brown, sulfur-containing animal pigments derived primarily from the oxidative cyclization of cysteinyldopa adducts formed by the coupling of L-cysteine to enzymatically generated dopaquinone (Prota, 1988). Typically, no indole-, pyrrole- or catechol-type products are formed upon degradation,

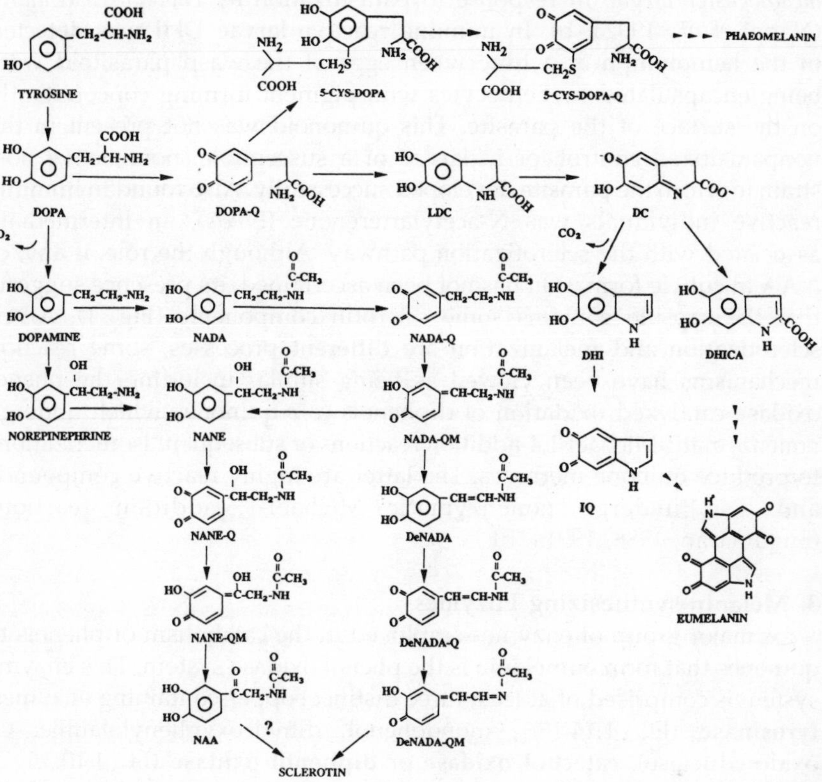

Fig. 1: The major components in the biosynthesis of eumelanin, phaeomelanin and sclerotin. DC, Dopachrome; DeNADA, dehydro-N-acetyldopamine; DeNADA-Q, dehydro-N-acetyldopamine quinone; DeNADA-QM, dehydro-N-acetyldopamine quinone methide; DHI, 5,6-dihydroxyindole; DHICA, 5,6-dihydroxyindole-2 carboxylic acid; DOPA, 3,4-dihydroxyphenylalanine; DOPA-Q, dopaquinone; LDC, leucodopachrome; NAA, N-acetylarterenone; NADA, N-acetyldopamine; NADA-Q, N-acetyldopamine quinone; NADA-QM, N-acetyldopamine quinone methide; NANE, N-acetylnorepinephrine; NANE-Q, N-acetylnorepinephrine quinone; NANE-QM, N-acetylnorepinephrine quinone methide; 5-CYS-DOPA, 5-cysteinyldopa; 5-CYS-DOPA-Q, 5-cysteinyldopaquinone; 5,6-INDO, 5,6-indole quinone.

but derivatives of thiazole- and pyridine-based acids are generated on permanganate oxidation. Allomelanins, or catechol melanins, are brown to black plant and fungal pigments derived from catechol or other simple polyphenols (Nicolaus, 1968). Upon alkali degradation allomelanins yield nitrogen-free compounds such as catechol, 1,8-dihydroxynaphthalene, and protocatechuic acid.

Positive identification of eumelanin in insect immunity has come from recent studies of alterations in hemolymph catecholamines of *Drosophila melanogaster* larvae in response to parasitization by *Leptopilina boulardi* (Nappi *et al.*, 1992a, b). In immune reactive larvae DHI was detected in the hemolymph at a time when eggs of the wasp parasitoid were being encapsulated by hemocytes with pigment forming concomitantly on the surface of the parasite. This quinonoid was not present in the nonparasitized control, or in larvae of a susceptible, nonreactive host strain in which the parasite developed successfully. Also found in immune reactive individuals was N-acetylarterenone (NAA), an intermediate associated with the sclerotization pathway. Although the role, if any, of NAA in cuticle formation has not been ascertained, its presence suggests that the capsule contains some sclerotin components (Fig. 1). Albeit sclerotization and melanization are different processes, some reaction mechanisms have been viewed as being similar including the phenol oxidase catalyzed oxidation of diphenols to *o*-quinones which undergo nonenzymatic Michael-1,4-addition reactions or subsequent isomerizations to produce quinone methides. The latter are highly reactive compounds and also undergo nonenzymatic Michael-1,6-addition reactions (Sugumaran, 1988, 1991a, b).

3. Melanin-synthesizing Enzymes

A major group of enzymes employed in the catabolism of phenols to quinones that form eumelanin is the phenol oxidase system. This enzyme system is comprised of at least three distinct copper-containing enzymes: **tyrosinase** (EC 1.14.18.1, monophenol, dihydroxyphenylalanine: O_2 oxidoreductase), **catechol oxidase** or **diphenol oxidase** (EC 1.10.3.1, o-diphenol: O_2 oxidoreductase), and **laccase** (EC 1.10.3.2, p-diphenol: O_2 oxidoreductase) (see review by Robb, 1984). Tyrosinase is unique among the three enzymes in that it catalyzes three different reactions in the biosynthetic pathway of eumelanin: (1) the initial hydroxylation of tyrosine to dopa (cresolase, hydroxylase, or monophenol oxidase activity); (2) the subsequent oxidation of dopa and a variety of other o-diphenols to their corresponding o-quinones (diphenol oxidase activity); and (3) the oxidation of DHI to indole-quinone (Hearing and Tsukamoto, 1991). Catecholase or diphenol oxidase is unable to catalyze the initial hydroxylation of tyrosine, but the enzyme is capable of initiating melanization by oxidizing o-diphenols to o-quinones. Neither tyrosinase nor catechol oxidase is capable of oxidizing p-diphenols, a function served only by laccase, which is also capable of oxidizing o-diphenols. Since laccase is presumed to be localized in the cuticle (Brunet, 1980), it follows that tyrosinase and catechol oxidase are the primary if not sole hemolymph phenol oxidases involved in the potentiation of melanotic encapsulations of intrahemocoelic parasites. Recent studies have shown enhanced monophenol and/or diphenol oxidase activities during insect cellular immune responses (Nappi *et al.*, 1987, Li *et al.*, 1989, 1992) but it remains to be determined whether these activities are generated by a single enzyme or if two enzymes are involved.

It should also be noted that other enzymes may rival tyrosinase in the early reaction sequence of melanogenesis. For example, iron-containing tyrosine 3-hydroxylase [L-tyrosine, tetrahydropteridine : oxygen oxidoreductase (3-hydroxylating) E.C. 1.14.16.2] can also hydroxylate tyrosine to dopa. However, this enzyme, which is localized in nervous tissue and requires the participation of a pteridine cofactor (2-amino-4-hydroxy-6,7-dimethyl-5,6,7,8-tetrahydropterine), may not be involved in the melanotic responses occurring in the hemolymph. Also, peroxidase (donor: hydrogen-peroxide oxidoreductase; EC 1.11.1.7), which is a heme-containing enzyme, can also catabolize phenols to o-diphenols and o-diphenols to o-quinones, making them available for melanin production. The possible involvement of cuticular laccase should not be discounted, since many parasites must penetrate the cuticular barrier to enter the hemocoel of their hosts.

The reaction sequence-forming eumelanin that follows the oxidation of dopa to dopaquinone includes intramolecular cyclization and indolization of dopaquinone to form leucochrome, dopachrome, DHI and/or DHICA, and 5,6-indole quinone. The widely held assumption that tyrosinase is the only enzyme involved in eumelanin biosynthesis was shown to be inaccurate by the recent discovery of dopachrome tautomerase, an enzyme which assists in the conversion of dopachrome to dihydroxyindoles (Pawelek *et al.*, 1980). This reaction is also nonenzymatically catalyzed by metal ions (Leonard *et al.*, 1988; Palumbo *et al.*, 1988, 1991). However, the enzyme seems to accelerate this step as well as confer a specificity to the reaction. Thus, dopachrome tautomerase isolated from mammalian systems seems to specifically catalyze the conversion of L-dopachrome to DHICA, while its counterpart in insects seems to generate DHI as the sole product (Aroca *et al.*, 1989, 1991; Aso *et al.*, 1989, 1990; Leonard *et al.*, 1988; Li and Nappi, 1991; Pawelek, 1990; Pawelek *et al.*, 1980; Sugumaran and Semensi, 1991; Hearing and Tsukamoto, 1991). The dihydroxyindoles thus formed can be further oxidized either enzymatically or nonenzymatically to produce indole quinones. The eventual nonenzymatic polymerization of indole quinones ultimately produces eumelanin. If the conversion of dopachrome is rate limiting, dopachrome tautomerase may play an important regulatory role in eumelanin biosynthesis during immune responses in insects. Dopachrome tautomerase has been isolated and/or characterized from different insects such as *Manduca sexta*, *Bombyx mori*, *Hyalophora cecropia*, and *Drosophila melanogaster*.

Enzymes that shift the metabolism of tyrosine and dopa away from eumelanin include dopa decarboxylase, dopamine β-hydroxylase, N-methyltransferase, and N-acetyltransferase. Oxidative deamination by monoamine oxidase, O-methylation by catechol-O-methyltransferase, and N-acetylation by N-acetyltransferase are possible reaction pathways for the degradation and inactivation of catecholamines and thus constitute important mechanisms for regulating catecholamine metabolism. The metabolism of β-hydroxylated phenylamines (such as norepinephrine) is primarily to the reduced state, while the metabolism of dopamine (and indolamines) is primarily to acid metabolites.

The metabolism of phaeomelanin arises by a deviation from the eumelanin pathway; the principal reaction involves the addition of cysteine to dopaquinone to produce 5-S-cysteinyldopa (Fig. 1) and, to a lesser extent, the 2-isomer (Prota, 1988). Subsequent oxidation yields the corresponding o-quinones which undergo cyclization to a thiazine ring. Reduction by cysteinyldopa provides a dihydrobenzothiazine that suffers oxidative coupling through an unknown mechanism to form the polymeric phaeomelanin. Apparently, there is no enzymatic activity beyond the formation of dopaquinone, a common precursor of both eumelanin and phaeomelanin. Consequently, the amount of available cysteine determines the type of melanin produced from dopaquinone and may explain the formation of melanin copolymers comprised of dopa and cysteinyldopa.

4. Factors Modulating the Formation of Eumelanin

Two reactions involving dopaquinone are of particular biological importance, the addition of sulfur nucleophiles and the rate of cyclization. The route taken by dopaquinone, either to form eumelanin or phaeomelanin, may be determined by the availability of free or conjugated sulfur. The cysteinyldopas which form phaeomelanins can be derived either by direct addition of cysteine to dopaquinone, or by enzymatic hydrolysis of glutathionyldopas. Low levels of glutathione reductase activity and corresponding low levels of cysteine-reduced glutathione (GSH) are associated with eumelanin formation, whereas high levels of enzyme activity are associated with phaeomelanin-producing melanocytes. Thus the thiol content and the activities of requisite enzymes may have a profound influence on the outcome of melanogenesis by determining whether eumelanin or phaeomelanin is produced. The nucleophilic addition of sulfur occurs preferentially at the 6-position of the dopaquinone ring rather than the 5-position, which is the common site of addition of nucleophiles. The regioselectivity of sulfur nucleophiles to the 6-position of dopaquinone profoundly affects the reactivity of the molecule which, once substituted, is not likely to undergo intramolecular cyclization to form eumelanin through the usual indole pathway.

The rate of cyclization of oxidized catechols is a major factor in determining the toxicity of catecholamines, with fast cycling N-methyl-substituted catecholamines being less toxic than unsubstituted ones (Crippa *et al.*, 1989). Dopaquinone undergoes rapid cyclization to form dopachrome. The subsequent reaction sequence leading to melanin may proceed spontaneously and involve two important indole intermediates, DHI and

DHICA. Under nonenzymatic conditions DHI predominates. The 5,6-indole quinone derived from DHI forms quinone-imine and quinone-methide tautomers which, if hydrated, form a transient trihydroxy indole (Lambert *et al.*, 1989). Dimerization and ensuing polymerization progress through a variety of condensation steps that may include the addition of methide to C4 of the trihydroxy indole. Since these reactions are generated spontaneously, considerable regulation would appear to be required to prevent unwarranted melanization. In this regard it is interesting to note that dopachrome is enzymatically converted by the activity of dopachrome tautomerase to DHICA. Unlike the autooxidation reaction of dopachrome which yields primarily DHI and DHICA, the enzymatic conversion of the iminochrome produces DHICA at the expense of the decarboxylated compound DHI. The latter reaction proceeds through a quinoine methide intermediate that rapidly tautomerizes to indole (Sugumaran 1991b; Sugumaran and Semansi, 1991). This sequence would appear to bypass DHI conversion to melanin precursors. Thus, it would appear that dopachrome tautomerase plays a major role in modulating the potentiation of melanin formation.

Despite a substantial amount of experimental work, the detailed mechanism of polymerization of dihydroxyindoles to form eumelanins remains obscure. A likely mechanism for the second-order decay of the radical derived from DHI is disproportionation giving rise to 5,6-indole quinone. Other possible products of the bimolecular decay of the radical derived from DHI or from the decay of 5,6-indole quinone are tautomeric quinone-imine and quinone-methide. The corresponding quinone-imine and/or quinone-methide tautomers derived from the decay of DHI and DHICA may then undergo nucleophilic addition of water, the extra OH being located at several possible positions, forming trihydroxyindoles. Such addition products could be short-lived, reacting as soon as they are formed with the remaining quinone-imine and quinone-methides forming dimeric structures (Lambert *et al.*, 1989).

5. Putative Killing Mechanisms Associated with Quinones

One important question that has never been adequately assessed concerns the biochemical components used by insects to kill the parasites they sequester in pigmented capsules. A variety of cytotoxic substances are likely to be produced by insects that incapacitate endoparasites, but this review will consider only those substances associated with eumelanin formation that may potentially target foreign surfaces for destruction.

It is well known that reactions involving tyrosinase generate reactive o-quinones as one of the major products (Riley, 1988; O'Brien, 1991). The initial reaction sequence in the biosynthesis of eumelanin includes two O_2-dependent steps: (1) o-hydroxylation of the monophenol L-tyrosine to form the o-diphenol L-dopa and (2) oxidation of L-dopa and certain other o-diphenols to form o-quinone. The reactions may be outlined as follows with RH_2 used to designate a cosubstrate:

(1) Monophenol + O_2 + RH_2 $\longrightarrow$ o-diphenol + H_2O + R
(2) o-diphenol + $1/2\ O_2$ $\longrightarrow$ o-quinone + H_2O

The metabolic fates of the various quinones generated during the oxidation of catechols by tyrosinase are not completely understood (O'Brien, 1991). The molecular basis of quinone cytotoxicity may be due to the covalent binding of quinone oxidation products with cell membrane components or other cellular nucleophiles, and/or the production of free radicals and reactive oxygen species from the univalent reduction of oxygen. Virtually overlooked among the many reactions occurring during enzyme-catalyzed melanotic encapsulation is the requirement for the reduction of O_2 to H_2O. For the initial hydroxylation catalyzed by tyrosinase, a cosubstrate (RH_2) furnishes two reducing equivalents to reduce one atom of molecular oxygen to water while the other atom is directly incorporated into the monophenol substrate. In the second reaction the four reducing equivalents required for the reduction of O_2 to H_2O are extracted from two diphenol substrates. The binuclear copper-containing enzyme tyrosinase can perform the tetravalent reduction of O_2 to H_2O ostensibly by transferring reducing equivalents in pairs. If the reduction of O_2 were to occur by successive univalent steps, highly reactive intermediate oxygen species would be produced, including superoxide anion $[O_2^-]$, hydrogen peroxide (H_2O_2), and the hydroxyl radical (.OH) (Halliwell and Gutteridge, 1989). Since these reactive oxygen species (ROS) are potentially deleterious to living systems, they are prime candidates for the cytotoxic molecules employed by insects to kill endoparasites. Although there may be sufficient evidence to implicate ROS in quinone toxicity, there is yet no experimental evidence that these reactive species are liberated at the active site of tyrosinase during melanotic encapsulations. The following considerations, based on known reaction mechanisms involving quinones and other eumelanin intermediates, are presented as a basis for future investigations aimed at identifying the molecular and biochemical mechanisms employed by insects to kill foreign organisms.

The first consideration is the mechanism(s) by which ROS might be generated during oxidation reactions leading to eumelanin formation. Although details of the polymerization process have not been fully elucidated, the principal derivatives of tyrosine incorporated into eumelanin include indole-5,6-quinone-2-carboxylic acid (IQCA), DHI and DHICA. The current view is that polymerization of these and other melanogenic intermediates occurs by free-radical reactions resulting from one-electron redox reactions which generate semiquinones (Riley, 1988).

The major species likely to be generated by one-electron exchange of either homologous or heterologous redox couples include dopa semiquinone, leucodopachrome semiquinone, and IQCA semiquinone (Fig. 2). Cytotoxic damage may involve sulphydryl oxidations, inactivation of DNA polymerase, depolymerization of lipids, and lipid peroxidation (Wick, 1980; Fitzgerald and Wick, 1983; Cadenas, 1989).

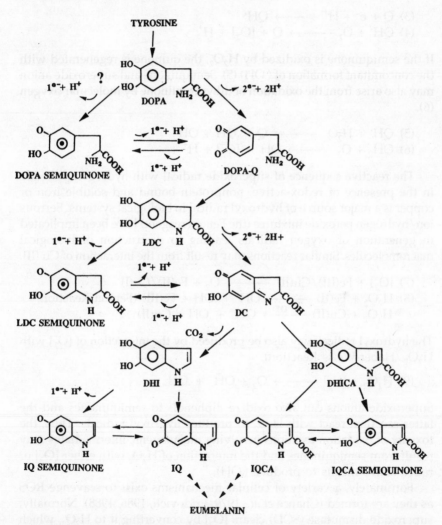

Fig. 2: One-electron transfers generate semiquinones which interact with molecular oxygen to produce reactive oxygen species in a process known as redox cycling. The alternative two-electron pathway may represent a protective mechanism against redox cycling by forming hydroquinones which are less reactive than semiquinones. DC, dopachrome; DHI, 5,6-dihydroxyindole; DHICA, 5,6-dihydroxyindole-2-carboxylic acid; DOPA-Q, dopaquinone; IQ, 5,6-indole quinone; IQCA, 5,6-indole quinone-2-carboxylic acid; LDC, leucodopachrome.

The mechanism believed to underlie the cytotoxicity of quinones is their propensity for redox-cycling, which can potentiate a cytotoxic free radical cascade generating reactive oxygen species. The process is initiated by a one-electron reduction of quinone (Q) to form a semiquinone (QH·) (3). The semiquinone is readily oxidized by O_2 with the concomitant formation of $[O_2^-]$ (4)

(3) $Q + e^- + H^+ \longrightarrow QH^·$

(4) $QH^· + O_2 \longrightarrow Q + [O_2^-] + H^+$

If the semiquinone is oxidized by H_2O_2, the quinone is regenerated with the concomitant formation of (.OH) (5). Semiquinone and superoxide anion may also arise from the oxidation of dihydroquinone by molecular oxygen (6).

(5) $QH^· + H_2O_2 \longrightarrow Q + .OH + OH^-$

(6) $QH_2 + O_2 \longrightarrow QH^· + [O_2^-] + H^+$

The reactive sequence of superoxide radical with hydrogen peroxide in the presence of redox-active, nonprotein bound and soluble iron or copper is a major source of hydroxyl radical in biological systems. Ferrous ion/hydrogen peroxide mixtures (the Fenton reagent) have been implicated in generation of oxygen radicals leading to destruction of biological macromolecules. Similar reactions may result from the interaction of Cu (II).

(7) $[O_2^-] + Fe(III)/Cu(II) \longrightarrow O_2 + Fe(II)/Cu(I)$

(8) $H_2O_2 + Fe(II) \longrightarrow OH^- + .OH + Fe(III)$ (Fenton Reaction)

$H_2O_2 + Cu(II) \xrightarrow{\text{or}} OH^- + .OH + Cu(II)$

The hydroxyl radical may also be produced by the interaction of $[O_2^-]$ with H_2O_2 (Haber-Weiss Reaction).

(9) $H_2O_2 + [O_2^-] \longrightarrow O_2 + OH^- + .OH$

Superoxide anions can also oxidize diphenols to semiquinones and the latter can then react with H_2O_2 to produce hydroxyl radicals. Thus, the toxicity of oxidation reactions involving melanogenic intermediates may result from semiquinones and the interaction of H_2O_2 with either $[O_2^-]$ or redox-active metals to produce (.OH).

Fortunately, a variety of cellular mechanisms exist to scavenge ROS as they are formed (Chance *et al.*, 1979; Fridovich, 1986, 1988). Normally, superoxide dismutase (SOD) clears $[O_2^-]$ by converting it to H_2O_2, which is then detoxified to H_2O by catalase and glutathione peroxidase (GP) (Fig. 3).

$$\text{SOD}$$
(10) $[O_2^-] + [O_2^-] + 2H+ \longrightarrow H_2O_2 + O_2$

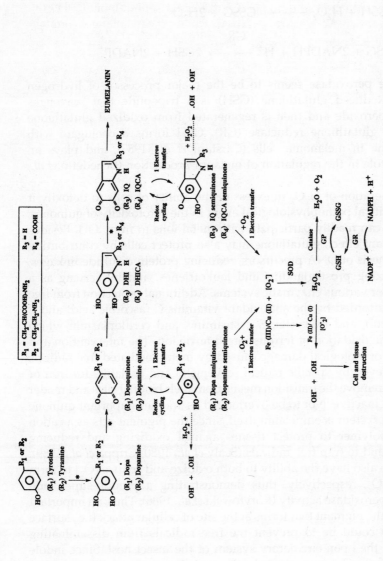

Fig. 3: The generation of free radicals and reactive oxygen species, and the mechanisms for detoxification. Superoxide dismutase (SOD) is a metaloenzyme that effectively removes superoxide anion, $[O_2^-]$, from the system by catalyzing its conversion to hydrogen peroxide (H_2O_2). Catalase and glutathione peroxidase convert H_2O_2 into water and the less active molecular oxygen (O_2). SOD may also play an important role in maintaining quinones in their fully reduced state as dihydroquinones, thus protecting against cyclized o-quinone toxicity. The toxicity of oxidation reactions is caused by hydroxyl-free radicals (.OH) generated from H_2O_2, and is directly related to the local concentration of Fe (II). Asterisk indicates reaction catalyzed by tyrosinase.

Catalase
(11) $2H_2O_2 \longrightarrow 2H_2O_2 + O_2$

GP
(12) $2 \, GSH + H_2O_2 \longrightarrow GSSG + 2H_2O$

GR
(13) $GSSG + 2NADPH + H^+ \longrightarrow 2GSH + 2NADP^+$

Glutathione peroxidase seems to be the major processor of hydrogen peroxide. Reduced glutathione (GSH) is a tripeptide that scavenges hydrogen peroxide and then is regenerated from oxidized glutathione (GSSG) by glutathione reductase (GR). GSH forms a conjugate with dopaquinone in melanoma cells (Carstam *et al.*, 1987), and plays an important role in the regulation of melanin production (Benedetto *et al.*, 1982).

The generation of H_2O_2 in excess of the ability of cells to detoxify it may be a critical pathophysiological event in the cytotoxicity of quinones, since H_2O_2 can readily participate with metal ions to from .OH. Besides its antioxidant effect, glutathione may also protect cells by contributing to the synthesis of DNA precursors, reducing protein disulfide linkages, biosynthesizing prostaglandins and leukotrienes, and by serving as a coenzyme for various enzymatic systems. Additional protection from free radicals is afforded by the antioxidant vitamins C (ascorbic acid) and E (α-tocopherol), and by cysteine, cysteamine, and ceruloplasmin which catalyzes the oxidation of ferrous ion to ferric ion. The intervention and prevention of biological damage caused by metal-mediated free radicals can be achieved by specific radical scavengers and chelators for iron or copper that remove the transition metals from their binding sites and render them redox inactive. Not to be overlooked as another important quinone detoxifying system is eumelanin itself, since the pigment acts as a cation exchange polymer to protect tissues against oxidizing and reducing conditions and to trap free radicals (Sealy *et al.*, 1980; Crippa *et al.*, 1989). Eumelanins also have the ability to both oxidize and reduce $[O_2^-]$ to form O_2 and H_2O_2, respectively, thus demonstrating a pseudo-superoxide dismutase-peroxidase activity (Korytowski *et al.*, 1986). Thus, an important function of the pigment that forms at the site of cellular attack (i.e., surface of parasite) could be to prevent the free radicals from disseminating throughout the open circulatory system of the insect host. Since indole derivatives formed enzymatically by the oxidation of certain catechols bind nucleophilic groups of proteins and/or chitin, it is possible that indole derivatives can also form covalent bonds with nucleophilic groups of biological membranes leading to degenerative changes and functional impairment of the cells. Repeated redox reactions might amplify the electron transfer to strong nucleophiles.

Hopefully, future studies will provide information as to whether these reactive species are generated in insects at the site of cellular interaction with foreign surfaces. It would be especially interesting to know what mechanisms exist for the site-specific localization of these components and for inactivating or sequestering these potentially damaging substances from affecting host tissues. Equally important information concerns the types of detoxification mechanisms insects possess that safeguard the integrity of the host and to what extent these mechanisms are modified during parasitization.

6. Phenoloxidase Activation

To understand the role played by quinones and other toxic melanogenic intermediates derived from the metabolism of tyrosine in insect immunity requires knowledge of the mechanisms of activation of phenoloxidase. Phenoloxidase occurs as an inactive proenzyme (prophenoloxidase) in the hemolymph of most of the insect species studied, including *Locusta migratoria* (Brehelin *et al.*, 1989), *Tenebrio molitor* (Heyneman, 1965), *Bombyx mori* (Ashida, 1971; Yoshida *et al.*, 1986; Ashida and Yoshida, 1988), *Antheraea pernyi* (Evans, 1967), *Manduca sexta* (Aso *et al.*, 1985), *Hyalophora cecropia* (Andersen *et al.*, 1989), *Galleria mellonella* (Pye, 1974), *Sacrophaga barbarta* (Hughes, 1976), *Calliphora vicina* (Thomson and Sin, 1970), *Musca domestica* (Tsukamoto *et al.*, 1986), *Drosophila melanogaster* (Seybold *et al.*, 1975), and *Aedes aegypti* (Ashida *et al.*, 1990). Activation of the proenzyme appears to be achieved by proteolytic cleavage of the proenzyme (Ashida and Dohke, 1980), and in some species divalent calcium ions appear to be required (Ashida and Yamazaki, 1990; Ashida and Yoshida, 1988). Several microbial products, such as bacterial, fungal and yeast cell wall components, cause prophenoloxidase activation in the hemolymph. The triggering substance from gram-positive bacteria seems to be peptidoglycan (Yoshida *et al.*, 1986; Brookman *et al.*, 1989).

In addition to protease-mediated activation, enzyme activation may result from other reagents, such as organic solvents, heavy metals, and heat, and surface active reagents, such as fatty acids and detergents. Recently, Sugumaran and Nellaiappan (1990, 1991) found that prophenoloxidase was specifically activated by treatment with detergents, such as sodium dodecyl sulfate, and low concentrations of phospholipids. Among the phospholipids tested, lysolecithin proved to be the most potent activator of lobster prophenoloxidase (Sugumaran and Nellaiappan, 1991). It appears that these exogenous substances bind to prophenoloxidase and alter the conformation of the enzyme, leading to exposure of its active site. In this regard, it is interesting to point out that during invasion, many researchers have demonstrated the presence of damaged hemocytes on or around the foreign objects (Ratcliffe *et al.*, 1985). Cellular damage can cause the release of phospholipase A_2 which, in turn, liberates lysolecithin and free fatty acids. The released lysolecithin, which is one of the most potent biological detergents, can function as a direct activator of phenoloxidase while the free fatty acids generated can be used for the synthesis of prostaglandins and eicosanoids, which could trigger further immune reactions as demonstrated recently (Blomquist *et al.*, 1991; Stanley-Samuelson *et al.*, 1991).

Since activated phenoloxidase is very sticky, it can adhere to foreign objects and other cell surfaces and generate cytotoxic quinonoid compounds to kill the intruder. Such toxic products of this enzyme activity are also dangerous for the host. Therefore, it is likely that the activated phenoloxidase is short-lived (Saul and Sugumaran, 1987) similar to some of the complement pathway components found in the blood of higher animals.

Since protease-mediated prophenoloxidase activation is well established, several workers have identified the presence of protease inhibitors that could control the prophenoloxidase activation in insects. Ramesh *et al.* (1988), Saul and Sugumaran (1986), and Sugumaran *et al.* (1985) have isolated three BPTI type serine protease inhibitors. Two inhibitors were isolated from *Manduca sexta* and another was isolated from *Sarcophaga bullata*. All three inhibitors drastically inhibit the protease-mediated prophenoloxidase activation. In *M. sexta*, one of the serine protease inhibitors could inhibit the cuticular protease-mediated activation of prophenoloxidase, confirming the proposed physiological role of this inhibitor (Saul and Sugumaran, 1986). Recently, Brehelin *et al.*, (1991) demonstrated the presence of a serine protease inhibitor that controls the prophenoloxidase activation in the hemolymph *Locusta migratoria*.

The speculation that prophenoloxidase systems could serve as a recognition mechanism of nonself-matter is an attractive but unsubstantiated one (Soderhall, 1982; Ratcliffe *et al.*, 1984; Leonard *et al.*, 1985). While the association of phenoloxidase with defence reactions of insects is well established, there is hardly any evidence in the literature to attribute a role for this enzyme in nonself-recognition. Although activation of the prophenoloxidase system in the wax moth *Galleria mellonella* by a microbial product enhanced the recognition of nonself-material (Leonard, *et al.*, 1985a; Ratcliffe *et al.*, 1984), this does not serve as proof for its role in nonself-recognition. Recently, Rizki and Rizki (1990) successfully demonstrated *Drosophila melanogaster* larvae lacking hemolymph phenoloxidase activity could readily encapsulate a foreign organism injected into the blood but could not melanize it. Therefore, the hypothesis that prophenoloxidase serves as a mechanism for recognition of foreignness should be reevaluated.

REFERENCES

Andersen, K., S.C. Sun, H.G. Boman and H. Steiner. 1989. Purification of the prophenoloxidase from *Hyalophora cecropia* and four proteins involved in its activation. *Insect Biochem.* 19: 629-637.

Aroca, P., F. Solano, J.C. Garcia-Borron, and J.A. Lozano. 1991. Specificity of dopachrome tautomerase and inhibition by carboxylated indoles. *Biochem. J.* 77: 393-397.

Aroca, P., J.C. Garcia-Borron, R.L. Solano, and J.A. Lozano, 1989. Regulation of mammalian melanogenesis. I: Partial purification and characterization of a dopachrome converting factor: dopachrome tautomerase. *Biochim. Biophys. Acta.* 1035: 266-275.

Ashida, M. 1971. Purification and characterization of prophenoloxidase from hemolymph of the silkworm *Bombyx mori*. *Arch. Biochem. Biophys.* 144: 749-762.

Ashida, M. and H.I. Yamazaki. 1990. Biochemistry of the phenoloxidase system in insects with special reference to its activation. In: *Molting and Metamorphosis*, pp. 239-265. (E. Ohnishi and H. Ishizaki, eds.). Japan Sci. Soc. Press, Tokyo/Springer-Verlag, Berlin.

Ashida, M. and H. Yoshida. 1988. Limited proteolysis of prophenoloxidase during activation by microbial products in insect plasma and effect of phenoloxidase on electrophoretic mobility of plasma proteins. *Insect Biochem.* 18: 11-19.

Ashida, M. and K. Dohke. 1980. Activation of prophenoloxidase by the activating enzyme of the silkworm, *Bombyx mori*. *Insect Biochem.* 10: 37-47.

Ashida, M., K. Kinoshita, and P.T. Brey. 1990. Studies on prophenoloxidase activation in the mosquito, *Aedes aegypti* L. *Eur. J. Biochem.* 188: 507-515.

Aso, Y., K.J. Kramer, T.L. Hopkins and G.L. Lookhart. 1985. Characterization of haemolymph protyrosinase and a cuticular activator from *Manduca sexta* (L). *Insect Biochem.* 15: 9-17.

Aso, Y., K. Nakashima and N. Yamasaki. 1990. Changes in the activity of dopa quinone imine conversion factor during the development of *Bombyx mori*. *Insect Biochem.* 20: 685-690.

Aso, Y., Y. Imamura and N. Yamasaki. 1989. Further studies on dopa quinone imine conversion factor from cuticles of *Manduca sexta*. *Insect Biochem.* 19: 401-407.

Beckage, N.E., J.S. Metcalf, D.J. Nesbit, K.W. Schleifer, S.R. Zetlan and I. deBuron. 1990. Host hemolymph monophenoloxidase activity in parasitized *Manduca sexta* larvae and evidence for inhibition by wasp polydnavirus. *Insect Biochem.* 20: 285-295.

Benedetto, J-P., J-P. Ortonne, C. Voulot, C. Kjatchadourian, G. Prota and J. Thivolet. 1982. Role of thiol compounds in mammalian melanin pigmentation. II. Glutathione and related enzymatic activities. *J. Invest. Dermatol.* 79: 422-424.

Blomquist, G.J., C.E. Borgeson and M. Vundla. 1991. Polyunsaturated fatty acids and eicosanoids in insects. *Insect Biochem.* 21: 99-106.

Boman, H.G., I. Faye, G.H. Gudmundsson, J.Y. Lee and D.A. Lidholm. 1991. Cell-free immunity in *Cecropia*. *Eur. J. Biochem* 201: 23-31.

Brehelin, M., L. Drif, L. Buad and N. Beomare. 1989. Insect hemolymph cooperation between humoral and cellular factors in *Locusta migratoria*. *Insect Biochem.* 19: 301-307.

Brehelin, M., R.A. Boigegrain, L. Drif and M.A. Colleti-Previero. 1991. Purification of a protease inhibitor which controls prophenoloxidase activation in hemolymph of *Locusta migratoria* (Insecta). *Biochem. Biophys. Res. Commun.* 179: 841-846.

Brookman, J.L., N.A. Ratcliffe and A.F. Rowley. 1989. Studies on the activation of the prophenoloxidase system of insects by bacterial cell wall components. *Insect Biochem.* 19: 47-57.

Brunet, P.C.J. 1980. The metabolism of aromatic amino acids concerned in the crosslinking of insect cuticle. *Insect Biochem.* 10: 467-500.

Cadenas, E. 1989. Biochemistry of oxygen toxicity. *Ann. Rev. Biochem.* 58: 79-110.

Carstam, R., C. Hansson, C. Lindbladhi, H. Rorsman and E. Rosengren. 1987. Dopaquinone addition products in cultured human melanoma cells. *Acta. Derm Venereol* (Stockholm) 67: 100-105.

Chance, B., H. Sies and A. Boveris. 1979. Hydroperoxide metabolism in mammalian organs. *Physiol. Rev.* 59: 527-605.

Christensen, B.M. and A.J. Nappi. 1988. Immune responses of arthropods. *ISI Atlas of Science. Animal Sciences*, pp. 15-19.

Coombe, D.R., P.L. Ey and C.R. Jenkin. 1984. Self and nonself recognition in invertebrates. *Q. Rev. Biol.* 59: 231-255.

Crippa, R., V. Horak, G. Prota, P. Svoronos and L. Wolfram. 1989. Chemistry of Melanins. In: *The Alkaloids*, vol. 36, pp. 253-323. Academic Press, New York.

Evans, J.J.T. 1967. The activation of prophenoloxidase during melanization of the pupal blood of the Chinese oak silkmoth *Antheraea pernyi*. *J. Insect Physiol.* 13: 1699-1711.

Fitzgerald, G.B. and M. Wick. 1983. 3,4-Dihydroxybenzylamine: An improved dopamine analog cytotoxic for melanoma cells in part through oxidation products inhibitory to DNA polymerase. *J. Invest. Dermatol.* 80: 119-123.

Fridovich, I. 1986. Biological effects of the superoxide radical. *Arch. Biochem. Biophys.* 247: 1-11.

Fridovich, I. 1988. The biology of oxygen radicals: General concepts. In: *Oxygen Radicals and Tissue Injury*. pp. 1-8. (B. Halliwell, ed.). Fed. Amer. Soc. Exper. Biol., Bethesda, MD.

Götz, P. 1986. Mechanisms of encapsulation in dipteran hosts. *Symp. Zool. Soc. Lond.* 56: 1-19.

Götz, P. and H.G. Boman, 1985. Insect immunity. In: *Comprehensive Insect Physiology, Biochemistry and Pharmacology* vol. 3, pp. 453-485. (G.A. Kerkut and L.I. Gilbert, eds.). Pergamon Press, Oxford.

Gupta, A.P. 1988. In: *Hemocytic and Humoral Immunity in Arthropods* (A.P. Gupta, ed.) John Wiley and Sons, New York.

Halliwell, B. and J.M.C. Gutteridge. 1989. *Free Radicals in Biology and Medicine*. Oxford Univ. Press, Oxford.

Hearing, V.J. and K. Tsukamoto. 1991. Enzymatic control of pigmentation in mammals. *FASEB J.* 5: 2902-2909.

Heyneman, R.A. 1965. Final purification of a latent phenoloxidase with mono- and diphenoloxidase activity from *Tenebrio molitor*. *Biochem. Biophys. Res. Commun.* 21: 162-169.

Hughes, L. 1976. Haemolymphal activation of protyrosinase and the site of synthesis of haemolymph protyrosinase in larvae of the fleshfly, *Sarcophaga barbarta*. *J. Insect Physiol.* 22: 1005-1011.

Kayser, H. 1985. Pigments. In: *Comprehensive Insect Physiology Biochemistry and Pharmacology*. vol. 10, pp. 367-415. (G.A. Kerkut and L.I. Gilbert, eds.). Pergamon Press, Oxford.

Korytowski, W., B. Kalyanaraman, I.A. Menton, T. Sarna and R.C. Sealy. 1986. Reaction of superoxide anions with melanins: electron spin resonance and spin trapping studies. *Biochim. Biophys. Acta* 882: 145-153.

Lackie, A.M. 1988. Haemocyte behaviour. *Adv. Insect Physiol.* 21: 85-178.

Lambert, C., J.N. Chacon, M.R. Chedekel, E.J. Land, P.A. Riley, A. Thompson and T.G. Truscott. 1989. A pulse radiolysis investigation of the oxidation of indolic melanin precursors: evidence for indole quinones and subsequent intermediates. *Biochimica et Biophysica Acta* 993: 12-20.

Leonard, C., N.A. Ratcliffe and A.F. Rowley. 1985. The role of prophenoloxidase activation in non-self recognition and phagocytosis by insect blood cells. *J. Insect Physiol.* 31: 789-799.

Leonard, L.J., D.W. Townsend and R.A. King. 1988. Function of dopachrome oxidoreductase and metal ions in dopachrome conversion in the eumelanin pathway. *Biochemistry* 27: 6156-6161.

Li, J. and A.J. Nappi. 1991. Electrochemical determination of dopachrome isomerase in larvae of *Drosophila melanogaster*. *Biochem. Biophys. Res. Comm.* 180: 724-729.

Li, J., J.W. Tracy and B.M. Christensen. 1989. Hemocyte monophenol oxidase activity in mosquitoes exposed to microfilariae of *Dirofilaria immitis*. *J. Parasitol.* 75: 1-5.

Li, J., J.W. Tracy and B.M. Christensen. 1992. Phenoloxidase activity in hemolymph compartments of *Aedes aegypti* during melanotic encapsulation reactiion against microfilariae. *Dev. Comp. Immunol.* 16: 41-48.

Nappi, A.J. and B.C. Christensen. 1987. Insect immunity and mechanisms of resistance by nematodes. In: *Vistas on Nematology* pp. 285-291. (J.A. Veech and D.W. Dickson, eds.). Society of Nematologists, Hyattsville, Maryland.

Nappi, A.J., B.M., Christensen and J.W. Tracy. 1987. Quantitative analysis of hemolymph monophenoloxidase activity in immune reactive *Aedes aegypti*. *Insect Biochem.* 17: 685-688.

Nappi, A.J., E. Vass, Y. Carton and F. Frey. 1992b. Identification of 3,4-dihydroxyphenylalanine, 5, 6-dihydroxyindole and N-acetylarterenone during eumelanin formation in immune reactive larvae of *Drosophila melanogaster*. *Archives of Insect Biochem.* 20: 181-191.

Nappi, A.J., Y. Carton and A.J. Frey, 1991. Parasite-induced enhancement of hemolymph tyrosinase activity in a selected immune reactive strain of *Drosophila melanogaster*. *Arch. Insect Biochem. Physiol.* 18: 159-168.

Nappi, A.J., Y. Carton, J. Li and E. Vass. 1992a. Reduced cellular immune competence of a temperature-sensitive dopa decarboxylase mutant strain of *Drosophila melanogaster* against the parasite *Leptopilina boulardi*. *Comp. Biochem. Physiol.* 101B: 453-460.

Nicolaus, R.A. 1968. *Melanins*. Hermann, Paris.

O'Brien, P.J. 1981. Molecular mechanisms of quinone cytotoxicity. *Chem. Biol. Interactions* 80: 1-41.

Palumbo, A., F. Solano, G. Misuraca, P. Aroca, J.C. Garcia Borron, J.A. Lozano and G. Prota. 1991. Comparative action of dopachrome tautomerase and metal ions on the rearrangement of dopachrome. *Biochimica et Biophysica Acta* 1115: 51-59.

Palumbo, A., M. D'Ischia, G. Misuraca, G. Prota and T.M. Schultz. 1988. Structural modifications in biosynthetic melanins induced by metal ions. *Biochimica et Biophysica Acta* 964: 193-199.

Pawelek, J.M. 1990. Dopachrome conversion factor functions as an isomerase. *Biochem. Biophys. Res. Commun.* 166: 1328-1333.

Pawelek, J., A. Korner, A. Bergstom and J. Bologna. 1980. New regulators of melanin biosynthesis and the autodestruction of melanoma cells. *Nature* 286: 617-619.

Prota, G. 1988. Progress in the chemistry of melanins and related metabolites. *Med. Res. Rev.* 8: 525-556.

Pye, A.E. 1974. Microbial activation of prophenoloxidase from immune insect larvae. *Nature* 251: 610-613.

Ramesh, N., M. Sugumaran and J.E. Mole. 1988. Purification and characterization of two trypsin inhibitors from the hemolymph of *Manduca sexta* larvae. *J. Biol. Chem.* 263: 11523-11527.

Ratcliffe, N.A., A.F. Rowley, S.W. Fitzgerald and C.P. Rhodes. 1985. Invertebrate immunity: Basic concepts and recent advances. *Int. Rev. Cytol.* 97: 183-325.

Ratcliffe, N.A., C. Leonard and A.F. Rowley. 1984. Prophenoloxidase activation: non-self recognition and cell cooperation in insect immunity. *Science* 226: 557-559.

Riley, P.A. 1988. Radicals in melanin biochemistry. *Annals New York Acad. Sci.* 551: 111-120.

Robb, D.A. 1984. Tyrosinase. In: *Copper Proteins and Copper Enzymes* vol. 2, pp. 207-240. (R. Lontie, ed.). CRC Press, Boca Raton, FL.

Salt, G. 1970. *The Cellular Defense Reactions of Insects*. Cambridge Univ. Press, Cambridge.

Saul, S.J. and M. Sugumaran. 1986. Protease inhibitor controls prophenoloxidase activation in *Manduca sexta*. *FEBS Lett.* 208: 113-116.

Saul, S.J. and M. Sugumaran. 1987. Protease mediated prophenoloxidase activation in the hemolymph of the tobacco hornworm *Manduca sexta*. *Arch. Insect Biochem Physiol.* 5: 1-11.

Schmidt, O. and U. Theopold. 1991. Immune defense and suppression in insects. *Bioessays.* 13: 343-346.

Sealy, R.C., C.C. Felix, J.S. Hyde and H.M. Swartz. 1980. Structure and reactivity of melanins: Influence of free radicals and metal ions. In: *Free Radicals in Biology* vol. 4., pp. 209-259. (W.A. Pryor, ed.). Academic Press, New York.

Seybold, W.D., P.S. Meltzer and H.K. Mitchell. 1975. Phenoloxidase activation in *Drosophila*: A cascade of reactions. *Biochem. Gen.* 13: 85-108.

Söderhall, K. 1982. Prophenoloxidase activating system and melanization. A recognition mechanism of Arthropods? A review. *Dev. Comp. Immunol.* 6: 601-611.

Stanley-Samuelson, D.W., E. Jensen, K.W. Nickerson, K. Tiebel, C.L. Ogg and R.W. Howard. 1991. Insect immune response to bacterial infection is mediated by eicosanoids. *Proc. Natl. Acad. Sci.* 88: 1064-1068.

Sugumaran, M. 1988. Molecular mechanisms for cuticular sclerotization. *Adv. Insect Physiol.* 21: 179-231.

Sugumaran, M. 1990. Prophenoloxidase activation and insect immunity. *UCLA Symposium on Mol. Cell. Biol. New Science* 121, 47 (1990).

Sugumaran, M. 1991a. Molecular mechanisms of sclerotization. In: *The Physiology of Insect Epidermis* pp. 141-168. (K. Binnington and A. Retnakaran, eds.). CSIRO Publications, Victoria.

Sugumaran, M. 1991b. Molecular mechanisms for mammalian melanogenesis. Comparison with insect cuticular sclerotization. *FEBS Lett.* 293: 4-10.

Sugumaran, M. and K. Nellaiappan. 1990. On the latency and nature of phenoloxidase present in the left colleterial gland of the cockroach *Periplaneta americana*. *Arch. Insect Biochem. Physiol.* 15: 165-181.

Sugumaran, M. and K. Nellaiappan. 1991. Lysolecithin—A potent activator of prophenoloxidase from the hemolymph of the Lobster, *Homarus americanas*. *Biochem. Biophys. Res. Commun.* 176: 1371-1376.

Sugumaran, M., S.J. Saul and N. Ramesh. 1985. Endogenous protease inhibitors prevent undesired activation of prophenolase in insect hemolymph. *Biochem. Biophys. Res. Commun.* 132: 1124-1129.

Sugumaran, M. and V. Semensi. 1991. Quinone methides as intermediates in eumelanin biosynthesis. *J. Biol. Chem.* 266: 6073-6078.

Thomson, J.A. and Y.T. Sin. 1970. The control of prophenoloxidase activation in larval haemolymph of *Calliphora*. *J. Insect Physiol.* 16: 2063-2074.

Tsukamoto, T., M. Ishiguro and M. Funatsu. 1986. Isolation of latent phenoloxidase from prepupae of the housefly, *Musca domesitca*. *Insect Biochem.* 16: 573-581.

Wick, M.M. 1980. Levodopa and dopamine analogs as DNA polymerase inhibitors and antitumor agents in human melanoma. *Cancer Res.* 40: 1414-1418.

Yoshida, H., M. Ochiai and M. Ashida. 1986. β-1,3-glucan receptor and peptidoglycan receptor are present as separate entities within insect prophenoloxidase activating system. *Biochem. Biophys. Res. Commun.* 141: 1177-1184.

CHAPTER 11

Haemagglutinins (Lectins) in Insects

J. P. N. Pathak

Introduction

Haemagglutinins or lectins are specific carbohydrate-binding proteins or glycoproteins and are present in most of the living organisms (Goldstein et al., 1980). The presence of natural haemagglutinins in the coelomic fluid of invertebrates was first discovered by Nogochi (1903). Since then, agglutinins have been reported in most of the advanced invertebrate phyla and reviewed by several workers (Cohen, 1970; Yeaton, 1981; Olafsen, 1986). Stein et al. (1932) posited that lectins which cause agglutination differ from stimulated agglutinating antibodies or immunoglobulins in structure, physico-chemical properties, origin and probably in function also. In many cases they resemble non-immunoglobulins, natural antibodies, and were previously classed as receptor-specific proteins (Gold and Balding, 1975). Yeaton (1981) considered that their universal occurrence indicated their existence as a kind of common molecule with some conventional function yet to be discovered. Among the invertebrate phyla, the molluscs have been well investigated for the presence of lectins and for their opsonic behaviour (Olafsen, 1986). In a series of papers, Renwrantz et al. (1978, 1981, 1983) demonstrated the opsonic behaviour of lectins in Helix pomatia and Mytilus edulus. Although in the last decade several researchers (Komano et al., 1981; Komano et al., 1983; Suzuki and Natori, 1983; Takahashi et al., 1985, 1986; Hapner and Stebbins, 1986; Stebbins and Hapner, 1986; Pendland and Boucias, 1985, 1988; Minuick et al., 1986; Bellah et al., 1988; Lackie and Vasta, 1988; Drif and Brehelin, 1989; Bradley et al., 1989; Pathak, 1991) have provided considerable information about the insect agglutinins, still, compared to molluscs, studies in insects on the lectins and their role in self and non-self discrimination require more attention from developmental and comparative immunologists. This paper reviews all the information available to date on the subject.

Methods of Study of Agglutinins

(a) *Selection of insects*: The availability of homogeneous insects is a main factor in most researches on insects. Each insect species must be maintained in the rearing room or insectory to make available insects of the same age, sex or physiological condition for assays of haemagglutinins.

(b) *Collection of haemolymph*: The availability of haemolymph varies in different species of insects due to their size and other factors. For example, enough haemolymph is available in lepidopteran larvae that a single insect can be used for this purpose while in other insects, such as hemipterans and coleopterans, the quantity of haemolymph is very little; hence their haemolymph is pooled in chilled aliquots from a fixed number of individuals. Sometimes, to obtain a sufficient quantity of haemolymph, insect blood is flushed from the insect by injecting a standard volume of a balanced saline solution (Lackie, 1981) into its abdomen. The collected haemolymph is centrifuged at 500-1000 g for 5 to 10 minutes at 4°C and the supernatant is stored at 4°C and used in various studies. Bellah *et al.* (1988) have suggested that for proper results samples must be assayed from many individuals at all stages and as early as possible after collection, as storage leads to impairment of agglutinating activity.

(c) *Preparation of erythrocytes, bacteria and fungi*: The erythrocytes of human and of other vertebrates are stored in Alsever's solution. The blood is usually discarded if not used within one week. The erythrocytes are washed three times in phosphate buffer solution (0.05 M KH_2PO_4, 0.1 M NaCl, pH 7.4). A 1% or 2% suspension of erythrocytes is used in agglutinin tests.

Bacterial or fungal spores are similarly prepared, after washing the culture in PBS and centrifuging the suspension at 1000 g for 10 minutes, it is suspended at a final concentration of 5×10^5 organisms/ml in PBS or any other concentration suitable to the experiment.

(d) *Haemagglutinin assays*: Agglutinin assays are performed in 'V'-bottomed wells in microtitre plates except in the case of rabbit erythrocytes, which give clear results on using 'U'-bottomed wells. One 100 μl of serial, twofold-diluted haemolymph in PBS and 100 μl of the erythrocyte, bacterial or fungal spore suspension is added to individual wells of microtitre plates. In the control, 100 μl of PBS and the same amount of erythrocyte suspension is used. The microtitre plate is maintained overnight at 4°C and the titres then recorded as the reciprocals of the highest dilutions at which haemagglutination occurred. Haemagglutination is indicated by the formation of a solid mat of erythrocytes while a dense dot (button) indicates that no agglutinins are present. A slight matting of erythrocytes is interpreted as trace agglutination.

Haemagglutination titres vary, depending on the carbohydrate binding preference of the agglutinin present in the haemolymph/tissue and on the type of RBC/micro-organism used in the essay. Sometimes the RBCs of one species are strongly agglutinated by the agglutinin present in the haemolymph while in other insects they may not be agglutinated by the particular agglutinin. This phenomenon is interpreted to mean that the susceptible erythrocyte or organism contains outer membrane glycoconjugates that are bound and cross-linked by the multivalent agglutinin, resulting in haemagglutination (Hapner and Stebbins, 1986). Several types of erythrocytes are generally used in screening for agglutinin activity in order to ensure its detection. Therefore, the assay results are always represented in different titre values against the type of erythrocytes involved in the particular study. Chemicals such as glutaraldehyde, formaldehyde, trypsin, protinase and neuraminidase etc., are capable of changing the chemical nature of the surface carbohydrate structures of RBCs and hence their treatment influences the titre value and/or specificity of erythrocyte agglutination, which enables exploration of the carbohydrate moiety of the relevant agglutinin.

Heteroagglutination detection can be performed as suggested by Scott (1971): 200 µl of haemolymph are treated with 50 µl erythrocytes or micro-organisms for 30 minutes at room temperature in an aliquot. The absorbing cells are removed by centrifugation at 500 to 1000 g and the remaining haemolymph is assayed for remaining agglutinin activity.

(e) *Inhibition of haemagglutinins*: The carbohbydrate binding specificity of haemagglutinins can be investigated through determination of agglutinin titre in the presence of potentially competing substances. A carbohydrate solution at an initial concentration of 200 mM in PBS is tested for its ability to inhibit the haemagglutinin titre. Serial dilutions are carried out in the sugar solutions using 50 µl of the diluted serum mixed well, the plates incubated for 1 hour at 4°C and then 50 µl of the erythrocyte micro-organism suspension added. A two- to fourfold decrease in titre value relative to the control is considered as inhibition of haemagglutination (Lackie, 1981; Jurenka et al., 1982; and others). A decrease in titre presumably means the inhibiting sugar is preferentially bound by the haemagglutinin molecule relative to erythrocyte surface receptors.

(f) *Isolation of agglutinin*: Standard biochemical methods are used for isolation of lectins from the haemolymph. Hapner and Jermyn (1981) employed the method of affinity chromatography, using small columns (0.9 × 5 cm) of Sepharose 4B-Fetuin and SDS polyacrylamide gel electrophoresis (7.5% gel) for purification of lectins. Other authors have used more or less similar methods with slight modifications.

Haemagglutinin Titres

Since the reports of Bernheimer (1952) and Feir and Walz (1964), a number of invertebrate immunologists have reported the occurrence of agglutinins in the haemolymph of insects (Donlon and Wemyss, 1976; Amirante 1976; Amirante and Mazzalai, 1978; Komano *et al.*, 1980; Hapner and Jermyn, 1981; Lackie, 1981; Pereira *et al.*, 1981; Jurenka *et al.*, 1982; Stynen *et al.*, 1982; Hapner, 1983; Ratcliffe and Rowley, 1983; Suzuki and Natori, 1983; Pathak, 1991). These workers have described the agglutinin titres against various vertebrate erythrocytes and micro-organisms in different species of insects.

It is evident from Table 1 that the agglutinin titre varies with the species, its developmental stage, age and sex, as well as the type of RBC used. No definite correlation has been found between species of insect and nature of RBC. Lackie (1981) reported that the serum of the cockroach possesses stronger agglutinins to a wide range of vertebrate erythrocytes than does the serum of *Schistocerca gregaria*. Hapner and Jermyn (1981) assayed the agglutinin activity in *Tellogryllus commodus* and found that there was no variation in the activity against different human blood groups (Table 1). The cells examined by them were grouped into three categories: strong (human, chicken, pigeon, rat) weak (cat and sheep) and no agglutinin activity (rabbit and horse). Hapner (1983) described the haemagglutinin titre in ten species of the family Acrididae, of which nine were grasshoppers of diverse geographic origin and one species was *Anabrus simplex* (cricket). The titre values varied from zero (no agglutination) to 256, which corresponded to 4096 for undiluted haemolymph. Haemolymph from individual insects showed variable agglutinating activity towards the same and different erythrocyte types. Hapner (1983) reported that human cell types O^+, B^+ and AB^+ gave higher titres compared to the erythrocytes of A^+ nature, which gave a lower titre in most of the grasshoppers tested. Jurenka *et al.* (1982) examined 17 species of Acrididae and noted that only 13 were able to agglutinate the erythrocytes. The authors concluded that haemagglutinin activity is present in the haemolymph of acridid insects but may vary in amount or even be absent in specific cases. They further concluded that the presence or absence of haemaglutinin in a species may reflect a selective physiological response towards environmental or biological factors. Thus Bellah *et al.* (1988) very rightly advised workers in insect immunity that to disprove absolutely the occurrence of agglutinin in an insect species, samples must be assayed for many individuals at all stages and as early as possible after haemolymph collection.

INDUCTION OF AGGLUTININ ACTIVITY

A few reports are also available on the induction of haemagglutinin activity in the haemolymph of insects after injury or infection. When the body wall of *S. peregrina* was injured by a hypodermic needle, the haemagglutinating activity was increased in the haemolymph (Komano *et al.*, 1980). Similar activity was also noted in pupal haemolymph 48 hours after puparium formation during the process of histogenesis; hence the authors correlated the occurrence of agglutinins with the process of non-specific immune surveillance. They further concluded that lectins probably help in scavenging the decomposed tissue fragments during the early pupal stage when larval tissues decompose and adult structures develop from imaginal discs. Kubo *et al.* (1984) also induced haemagglutinating activity in adults of *Sarcophaga peregrina* when the body wall was injured with a hypodermic needle. The nature of the lectin induced in the adults was the same as that found in the haemolymph of injured larvae; however, the degree of activity induced was less in adults. Using sugar specificity and cross-reactivity tests Kubo's team proved that the nature of the lectin molecule was the same as that of the lectin purified from injured larvae. Pendland and Boucias (1985) demonstrated the induction of haemagglutinin activity in the haemolymph of *A. gemmatalis* larvae by intrahaemocoelomic injection of a fungal cell suspension. The activity titres increased at 24 and 48 hours post-injection of (FL78-6W) hyphal bodies. These studies suggest that in some insects agglutinin activity may be induced by injecting the pathogens or by trauma.

Haemagglutinins in Relation to Development and Age

Suzuki and Natori (1983) studied the changes in haemagglutinating activity in the haemolymph of *B. mori* during development. They found that the haemagglutinating titre was relatively high in the early fifth instar, during the spinning period and at the emergence of adult and decreased during intervening periods. Thus the activity was the highest at the prepupal stage with no haemagglutinin activity in the haemolymph at the pupal stage. The haemagglutinin activity was again restored in the adult. The authors suggested that these changes are essential for the development of the insect as the agglutinins play a role in scavenging the decomposed self-tissue fragments during metamorphosis. Komano *et al.* (1980, 1981) have proposed a similar role for the lectins (agglutinins) in *Sarcophaga peregrina* larvae during metamorphosis. They observed that in the flesh-fly, *S. peregrina*, larval haemolymph contains very low amounts of active agglutinin, which is induced to higher levels in response to injury of the body wall or upon pupation.

Pathak (1991) indicated a possible correlation between lectin and metamorphosis but could provide no direct evidence to prove their role in self-non-self differentiation. He noted a cyclical change in the haemagglutinin activity of *Halys dentata* during development and reported that agglutinin titres were weak in two-day-old 3rd instar nymphs against various erythrocytes and micro-organisms (Fig. 1). The titres increased with age. In the 3rd instar nymphs the titre was highest on the 6th day against

Table 1: Haemagglutinin titres against erythrocytes of vertebrates in different species of insects

S.No.	Author	Insect species	Stage	Erythrocytes	Titre^{-1}
1.	Lackie (1981)	*Periplaneta americana*	Adult	Horse	1024-4096
				Rat	512
				Rabbit	4096
				Sheep	256
				Human 'O'	4096
2.	Lackie (1981)	*S. gregaria*	Adult	Horse	Zero
				Rat	64-128
				Rabbit	512
3.	Hapner and Jermyn (1981)	*Teleogryllus commondus*	Adult	Human (different types)	32
				Pigeon	64
				Chicken	32
				Rat	32
				Cat	04
				Rabbit	Not agglutinated
				Rabbit (Trypsin treated)	16
				Horse	Not agglutinated
				Monkey	02
				Sheep (Trypsin treated)	02
				Sheep (Trypsin treated)	4-8
4.	Pereira *et al.* (1981)	*Rhodnius prolixus*	Larvae/adult	Rabbit	08
				Dog	04
5.	Stynen *et al.* (1982)	*Leptinotarsa decomlineata*	Larva (last stage)	Human A$^+$	128
				Human A$^-$	256
				Human B$^+$	128
				Human AB$^-$	128
				Human O$^+$	128

No.	Reference	Species	Stage	Cell tested	Titre
6.	Jurenka *et al.* (1982)	*Melanoplus sanguinipes*	Adult	Human O⁻	128
				Rabbit	256
				Ox	256
				Rat	512
				Horse	3200
				Rabbit	256
				Human (all types)	32
				Calf	32
				Guinea pig	32
				Cat	32
				Pig	32
				Mouse	16
				Chicken	04
				Sheep	04
				Nosema locustae (spores)	08
7.	Ratcliffe and Rowley (1983)	*Anax imperator*	Larva	Sheep	16
8.	Pendland and Boucias (1985)	*Anticarsia gemmatalis*	Larva	Rabbit	Trace
		Trichoplusia ni	Larva	Sheep	256-512
				Human 'O'	Zero
				Rabbit	02
				Sheep	02
				Human 'O'	02
		Spodoptera frugiperda	Larva	Rabbit	04
				Sheep	02
				Human 'O'	
9.	Bellah *et al.* (1988)	*Philosamia ricini*	Fifth instar larva	Human 'O'	0-9 (log 2)
				Guinea pig	
				Rat	

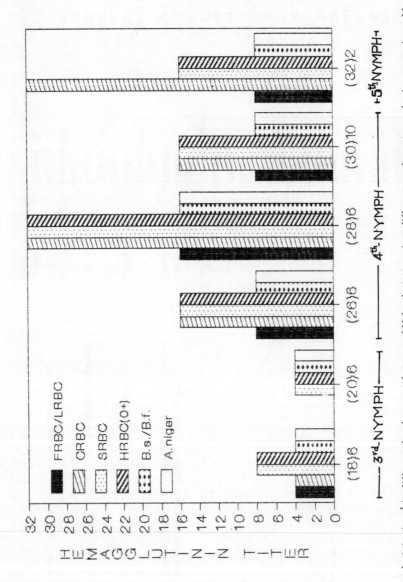

Fig. 1: Haemagglutinin titre⁻¹ in different developmental stages of *Halys dentata* against different erythrocytes and micro-organisms. Number in parentheses represents the age of insect after hatching. Reprinted with permission from *Developmental and Comparative Immunology*, vol. 15 (1991). Pathak, J.P.N. Occurrence of natural hemagglutinins in *Halys dentata* (Hemiptera) during development.

sheep erythrocytes (SRBC) and human O⁺ erythrocytes (HRBC O⁺). The activity declined on the 8th day before ecdysis. The titre remained nearly constant in the early 4th instar after ecdysis, peaked in 8 days and again reduced on the 9th day before ecdysis. In 5th instar nymphs the highest activity was noted on the 10th day, declining again on the 14th day before the final moult. In adults also the activity was initially low but increased on the 6th day and remained constant thereafter (Fig. 2). Thus Pathak (1991) concluded that in *Halys dentata* haemagglutinin activity varies at each stage, i.e., from apolysis to apolysis. Activity decreased prior to ecdysis and increased in the intervening period. This cyclical variation in the agglutinin titre in *H. dentata* cannot be explained on the basis of the concept proposed by Komano *et al.* (1980, 1981) because in a hemimetabolic insect (*H. dentata*) the apolysis-ecdysis phase is very short and restricted to the formation of new cuticle, which is dependent on the quantity of juvenile hormone in the blood (Wigglesworth, 1973). Hence Pathak suggested that in future, hormonal titres in haemolymph might be assayed to establish a direct relationship between lectins, metamorphosis and the immunosurveillance mechanism in insects. With the help of immunoelectrophoresis, Stynen *et al.* (1982) reported two different agglutinins in the Colorado beetle (*Leptinotersa decemlineata*) during development. They characterized one as a larval-pupal haemagglutinin as it was specifically present in larvae and pupae and identified the second as chromoprotein 2, which was present in all stages including eggs and adults.

Variation in the haemagglutinin titre with age of the insect has been well studied in the last instar larva of *Philosamia ricini* (Bellah *et al.*, 1988). The authors found no agglutinin in the haemolymph of one to four-day-old larvae while the levels of agglutinin against HRBC peaked in nine-day-old larvae and dropped sharply on the 11th and 12th day. The agglutinin titre against GRBC increased enormously on day 5 and varied between 6-9 (log 2) until the prespinning stage. However, the titre against *Bacillus thuringiensis* was nearly constant at all days of age. Bellah and colleagues studied the variation in agglutining titres in two different seasons, summer and winter, and found that it varied with the age of larvae.

Haemagglutinins in Relation to Sex

Sex had no specific impact on the haemagglutin titre in most insects. However, in grasshoppers (Jurenka *et al.*, 1982), females showed a slightly higher activity compared to males. But in *Blaberus craniifer* (Donlon and Wemyss, 1976) and *T. commodos* (Hapner and Jermyn, 1981) the activity was similar in both sexes. In *Philosamia ricini*, titres of agglutinins in males and females were highly comparable (Bellah *et al.*, 1988). Similarly in *Halys dentata* (Pathak, 1991) sex exerted no influence on the agglutinin titres. Apparently the agglutinin titre in insects is not influenced by sex.

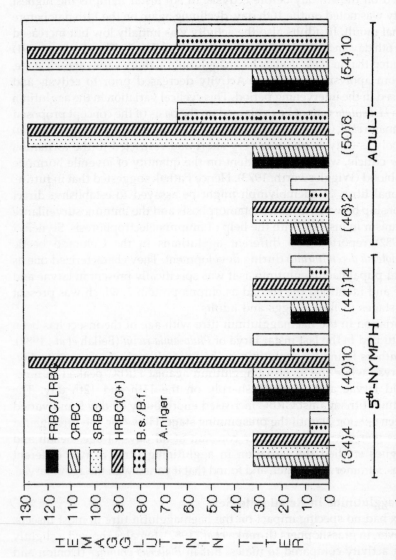

Fig. 2: Haemagglutinin titre^{-1} in different developmental stages of *Halys dentata* against different erythrocytes and micro-organisms. Number in parentheses represents the age of insect after hatching. Reprinted with permission from *Developmental and Comparative Immunology*, vol. 15 (1991). Pathak, J.P.N. Occurrence of natural hemagglutinins in *Halys dentata* (Hemiptera) during development.

Heterogeneous Nature of Agglutinins

Absorption studies have indicated the presence of different types of agglutinins in the haemolymph of the same species. Lackie (1981) noted that preincubation of cockroach serum with rabbit erythrocytes reduced the agglutination titre towards SRBC while preabsorption with horse and rat RBC had no or little effect upon subsequent agglutination of SRBC. She further suggested that the specificity of the agglutinin molecule which agglutinates rat and horse RBC is very different from that which reacts with the RBC of rabbit and sheep. Similarly Jurenka *et al.* (1982) decreased the heteroagglutinin activity towards distinct erythrocytes in *M. sanguinipes*. They found that preabsorption with human RBC removed haemagglutinin activity towards human and cat RBC, but could not reduce the activity titre against calf or rabbit erythrocytes, while preabsorption with rabbit erythrocytes removed activity of all the erythrocytes tested by them. This suggested that the agglutinin responsible for the agglutination of rabbit and calf erythrocytes is distinct from that which agglutinates human and cat red cells. Jurenka and co-workers further suggested that the red blood cells of rabbit were able to absorb all of the heteroagglutinins from the haemolymph of *M. sanguinipes*. Similar results were also observed by Scott (1971) and Lackie (1981) in *Periplaneta americana*. Pendland and Boucias (1985) have performed a comparative study on the haemagglutinin activity of some lepidopteran larvae and found that *Anticarsia gemmatalis* larvae also contain heteroagglutinins.

Pathak (1991) assayed the heteroagglutinins in eight-day-old *Halys dentata*. As shown in Fig. 3, the preabsorption of haemolymph with specific erythrocytes or micro-organisms decreased or removed the agglutinin. Preabsorption by either erythrocytes of frog or erythrocytes of lizard (LRBC) removed the agglutinating activity against both cell types, so they were probably agglutinated by the same agglutinin. Preabsorption by chick erythrocytes (CRBC) reduced the agglutinin titre against FRBC and LRBC, leaving traces of agglutination activity. CRBC absorption also reduced the agglutination titre against other erythrocytes. Similarly, preabsorption with sheep erythrocytes (SRBC) removed the agglutinin for CRBC and SRBC; it also reduced the titres in other cases. However, preabsorption with erythrocytes exerted very little influence on agglutination titres against micro-organisms or vice versa. He further concluded the presence of two different agglutinins as preabsorption of haemolymph with *Aspergillus niger* removed the specific agglutinin while preabsorption with *Bacillus thuringiensis* was capable of removing both the agglutinin which reacts against bacteria and that which reacts against fungal spores. These studies indicate that the heteroagglutinins seem to occur ubiquitously in insects.

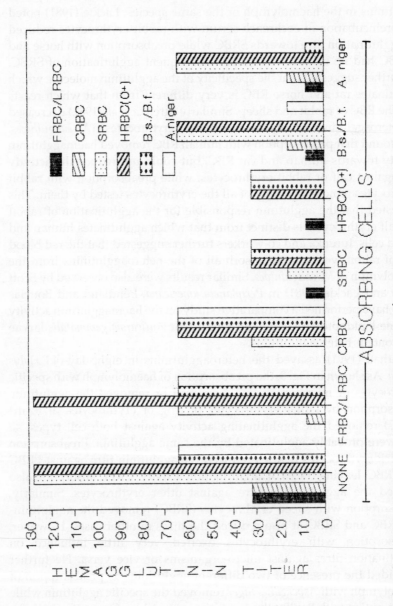

Fig. 3: Effect of preabsorption on naturally occurring haemagglutinin titre⁻¹ in eight-day-old adult *Halys dentata* against different erythrocytes and micro-organisms. Reprinted with permission from *Developmental and Comparative Immunology,* vol. 15 (1991). Pathak, J.P.N. Occurrence of natural hemagglutinins in *Halys dentata* (Hemiptera) during development.

Inhibition of Haemagglutinins

Haemagglutinins are specific carbohydrate binding proteins. A wide range of sugar and glycoproteins have been tested by various workers to establish the inhibitory and non-inhibitory carbohydrates. No general pattern can be presumed for the binding preference of various sugars in different insect species. Table 2 lists the various inhibiting substances used by some workers in inhibiting the agglutinins in different species of insects. The complex inhibitory activities suggest the presence of heteroagglutinins in most insects (Lackie, 1981). As there is no similarity in the inhibition pattern of two species, it is believed that the sugar specificity and the inhibition pattern may be the distinctively characteristic of individual species (Jurenka *et al.*, 1982). Stebbins and Hapner (1985) purified the agglutinin from grasshopper haemolymph and noted that its agglutination nature towards human erythrocytes was inhibited by D-galactoside and D-glucosidic sugars as well as by EDTA. This inhibitory behviour indicates the metallo-protein and specific binding nature of haemagglutinins.

Recently Mckenzie and Preston (1992a) isolated a lectin from the haemolymph of the blowfly larvae *Calliphora vomitoria* which agglutinated the mammalian erythrocytes with varying specificities and was inhibited by D-galactose and fetuin.

Site of Synthesis of Haemagglutinins

Most of the reports concerning haemagglutinins in insects describe the natural agglutination activity of haemolymph. However, a few for example Mauchamp (1982), describe an agglutinin as present in the epidermal cell membranes of newly apolysed pupal wings of the cabbage butterfly, *Pieris brassicae*, while the haemolymph and haemocyte extracts showed no haemagglutination activity. Similarly, in cockroaches an agglutinin has been reported from the membrane fraction of homogenized coxal depressor muscles (Denburg, 1980). The flesh-fly *Sarcophaga peregrina* contains an agglutinin in both the haemolymph as well as in the fat body (Komano *et al.*, 1980, 1983). The larval haemolymph from *S. peregrina* normally contains a very low amount of an active agglutinin, which is induced to higher levels in response to injury of the larval body wall or upon pupation (Komano *et al.*, 1980). Komano *et al.* (1983) confirmed that the agglutinin present in *S. peregrina* is synthesized in the fat body and released into the haemolymph upon injury to the larval body wall and also during pupation. The agglutinin is then apparently recruited by the haemocytes. However, Takahashi *et al.* (1984) could not demonstrate the synthesis of agglutinin in isolated fat bodies with the help of radiolabelled substances in *Sarcophaga peregrina* larvae.

A meticulous effort was made by Stiles *et al.* (1988) to demonstrate the site of synthesis of the haemolymph agglutinin utilizing primary cultures of several tissues from *M. differentialis* in combination with metabolic

Table 2: Haemagglutinin inhibitory substances

S. No.	Species	Inhibitory substances
1.	*S. peregrina* (Komano *et al.*, 1980)	D-galactose Lactose
2.	*T. commodus* (Hapner and Jermyn, 1981)	N-acetyl-D-neuraminic acid, N-acetyl-D-glucosamine, N-acetyl-D-galactosamine, Fetuin, EDTA
3.	*Leptinotarsa decemlineata* (Stynen *et al.*, 1982)	D-glucosamine, D-galactosamine, D-manmosamine, D-glucose-6-phosphate, Heparin, Dextransulphate, Mucin
4.	*M. sanguinipes* (Jurenka *et al.*, 1982)	L-arabinose, D-arabinose, D-fucose, L-rhamnose, D-xylose, D-glucose, D-galactose, N-acetyl-D-galactosamine, Lactose, Melibiose, maltose, sucrose, Stachyose, Methyl-D-glucopyramoside Fetuin, thyroglobulin, Bovine mucin, Fibrinogen, Ovalbumin

radiolabelling of synthesized proteins. By using the ELISA technique they found that haemocytes appeared to release only a minimal amount of agglutinin into the medium while the highest release was noted from fat bodies. The authors further noted that the small amount of radiolabelled agglutinin released by the ovary and testis cultures probably originated from the highly active contaminating fat body cells with these tissues. They also induced the fat body culture by microbial surface or cell wall components but they could not induce an increased agglutinin release. However, laminarin stimulated an increase in the rate of agglutinin release compared to control cultures. But Stiles and his co-workers concluded that this effect was transistory and, furthermore, that an increased rate of release of agglutinin from a treated culture could not prove the induction of synthesis. Such could be considered only a modulation of the rate of release from the tissue. Thus, except in *S. peregrina*, the site of agglutinin synthesis has not yet been determined though in most insects the fat bodies have been suggested as the site (Amirante and Mazzalai, 1978; Yeaton, 1981; Komano *et al.*, 1983). Hapner and Stebbins (1986) have pointed out that it is not known whether certain haemocytes are involved in the sequestration of the agglutinin. It is also not confirmed that the active agglutinin is synthesized *de novo* by the fat body or the pre-existing inactive agglutinin is withdrawn from the haemolymph into the fat body for activation and then returned to the haemolymph.

Chemical Nature of Haemagglutinin

Studies on the biochemical nature of agglutinin in insects are limited. Hapner and Jermyn (1981) isolated the agglutinin of *Teleogryllus commodus*, which has a larger molecular weight (> 1,000,000). The subunits of this protein have a molecular weight of 53,000 and 31,000. In *Sarcophaga peregrina* (Komano *et al.*, 1980) the active lectin was found to have a molecular weight of 190,000 and to consist of four α subunits and two β subunits, with molecular weights of 32,000 and 30,000 respectively. Suzuki and Natori (1983) isolated a component with haemagglutinin activity from the haemolymph of silkworm larvae. The molecular weight of the partially purified protein was 260,000. Stebbins and Hapner (1985) analyzed the amino acid composition of purified haemagglutinin from *M. sanguinipes* and *M. differentialis*. They reported all the usual amino acid residues, with the amounts of aspartic acid and glutamic acid being the highest and methionine the least. Hapner and Stebbins (1986) reviewed the biochemical properties of arthropod agglutinins for a few insect species. They mentioned that most of the agglutinins are high molecular weight, multimeric structures composed of low molecular weight protein subunits (Table 3). They further described that the quaternary structure of such multimers is generally stabilized by non-covalent forces, but the contributory subunits consist of smaller polypeptide chains that may be stabilized by inter- or intradisulphide bonds.

Opsonic Properties of Haemagglutinins

Induction of agglutinins by injury or on introduction of a micro-organism into the haemocoel of the flesh-fly, *Sarcophaga peregrina* (Komano *et al.*, 1981; Komano *et al.*, 1983; Takahashi *et al.*, 1985; 1986) and *Anticarsia gemmatalis* (Pendland and Boucias, 1985) indicated their probable role in insect immunity, specifically in binding or phagocytic activity. The carbohydrate binding capacity of insect agglutinins also signifies their proposed participation in early recognition steps in pathogen clearance (Sharon, 1984; Gupta, 1986). Agglutinins may serve as a discriminatory link between non-self object and the haemocytes involved in phagocytosis or encapsulation reaction. Ratner and Vinson (1983) proposed that haemocyte membrane-bound haemagglutinin serves to directly bind foreign particles, and humoral haemagglutinins serve as opsonic factors that bind to foreign particles and facilitate their uptake by haemocytes. This hypothesis has not been proved by any worker in insects; however, agglutinins have been shown to act as opsonins in molluscs and to aid in phagocytosis (Renwrantz *et al.*, 1981; Renwrantz and Stahmer, 1983).

The presence of haemolymph opsonins in insects is not proved. However, in *Spodoptera exigna* haemolymph lectin did show opsonization of fungal cells (Pendland and Boucias 1988). Bradley *et al.* (1989) demonstrated the haemocytic location of grasshopper agglutinin by immunocytofluorescence techniques and also studied the opsonic activity of the purified molecule. They purified and isolated a 95% pure haemagglutinin from the haemolymph of *M. differentialis*. The molecular

Table 3: Molecular weight of purified agglutinin protein from different species of insects (modified and adopted from Hapner and Stebbins, 1986)

S.No.	Insect species	Molecular weight or Native Molecular weight	Subunit MW	Author
1.	*Sarcophaga peregrina*	190,000	32,000, 30,000	Komano et al. (1980)
2.	*Teleogryllus commodus*	71,000,000	53,000 and 31,000	Hapner and Jermyn (1981)
3.	*Pieris brassicae*	43,000	23,000	Mauchamp (1982)
4.	*B. mori*	260,000	ND	Suzuki and Natori (1983)
5.	*Allomyrina dichotoma*			
	A-I	65,000	17,500, 20,000	Umetsu et al. (1984)
	A-II	66,500	19,000, 20,000	
6.	*Melanoplus differentialis*	500,000-700,000	70,000, 40,000, 28,000	Stebbins and Hapner (1985)
7.	*Melanoplus sanguinipes*	500,000-700,000	70,000, 40,000, 28,000	Stebbins and Hapner (1985)

ND = Not determined

weight of the purified agglutinin was 61,000. The serum or isolated agglutinin had no effect *in vitro* on haemocytic adherence or phagocytic activity. The author also noted that prior incubation of the monolayer of haemocytes with monoclonal antibody had no effect on the phagocytic behaviour of the plasmatocytes. Furthermore, their studies could not demonstrate any opsonic involvement for grasshopper serum or purified agglutinin towards human RBC, *B. thuringiensis* or *N. locustae* in a haemocyte monolayer system. Therefore, Bradley and co-authors concluded that in the grasshopper, agglutinin is not generally involved in an opsonic action or as a recognition molecule in the interaction between haemocytes and foreign particles.

Drif and Brehelin (1989) studied the clearance of sheep RBC and rabbit RBC from the haemolymphs of *Locusta migratoria* and found that rabbit RBCs were cleared much faster than SRBCs. In *in vitro* studies they found that rabbit RBC readily attached to haemocytes and numerous 'rosettes' were found with locust haemocytes. The SRBCs were unable to form 'rosettes' with haemocytes. This reaction of locust haemocytes was interpreted as the evidence for the opsonic action of the agglutinin present in the plasma or the serum. They further supported their conclusion with *in vitro* experiments in the presence of an agglutinin inhibitor, which reduced the adherence percentage of locust haemocytes with rabbit RBC. That agglutinin acts as a recognition factor was also the conclusion drawn from studies by Takahashi *et al.* (1984) in *Sarcophaga peregrina*. They showed that a specific agglutinin appears in the haemolymph of the pupa which assists in clearing histolysing tissue fragments. Similar conclusions were also drawn by Pendland and Boucias (1988) and Lackie and Vasta (1988) for *Spodoptera exigna* and *P. americana* respectively. Recently McKenzie and Preston (1992a) isolated and purified the lectin from haemolymph of *Calliphora vomitoria*. They examined the opsonic behaviour of purified lectin and the data are awaited (McKenzie and Preston, 1992b). However, Ratcliffe and Rowley (1983) denied the presence of an opsonic substance in the haemolymph of insects.

Hapner (personal communication, 1991) affirmed the opsonic activity of grasshopper agglutinin towards *Beauvaria bassiana* (fungal) blastospores. He wrote:

> We have shown that purified agglutinin increases the association of blastospores and hemocyte monolayers by four-to-sixfold. The opsonic stimulation occurs both when blastospores or the hemocyte monolayer is initially treated with agglutinin. This suggests that the agglutinin forms a bridge between the hemocyte and the fungal cell. No opsonization is observed with *Nomuraea rileyi*, a fungus that lacks surface galactose residues. The opsonization process is inhibited by competing carbohydrates, EDTA and agglutinin specific antibodies. *In vivo* experiments show that *B. bassiana* blastospores treated with

agglutinin are cleared from the hemolymph faster than those not treated, indicating a possible physiological functional activity of the grasshopper agglutinin. Our previous work (Bradley *et al.*, 1989) showed that the agglutinin did not opsonize bacteria, RBC or protozoan spores. Therefore, the agglutinin likely conveys opsonic activity only in selected cases.

Conclusions

Haemagglutinins or lectins are specific carbohydrate binding non-enzymatic proteins or glycoproteins of ubiquitous occurrence. They bind to erythrocytes and other cell surfaces having glycoconjugants to agglutinate or precipitate them. The agglutinating capacity of a lectin is inhibited by certain sugars. A decrease in the haemagglutinin titre presumably means that the inhibiting sugar is preferentially bound by the haemagglutinin molecule in relation to erythrocyte surface receptors. Haemagglutinins may be undetectable in certain species, tissues or developmental stages. The sites of biosynthesis of these molecules are not yet known, except for *S. peregrina* in which synthesis takes place in the fat bodies (Komano *et al.*, 1983).

Several agglutinins have been purified from different insect species. Their molecular weight was high and the content of aspartic acid and glutamic acid maximum with methionine least. The haemagglutinin titre may vary in different developmental stages or with the age of an insect. A cyclical change related to metamorphosis has also been demonstrated. These observations would seem to indicate opsonic behaviour or a scavenging role for haemagglutinins during histogenesis. Though it is knwon that a purified agglutinin increases the association of blastospores and haemocyte monolayers, some additional direct evidence must be adduced to prove that the agglutinins do indeed play a role in self and non-self discrimination among insects, i.e., in insect immunity.

REFERENCES

Amirante, G.A. 1976. Production heteroagglutinins in haemocytes of *Leucophea maderae* L. *Experientia* 32: 526-528.

Amirante, G.A. and F.G. Mazzalia. 1978. Synthesis and localization of hemagglutinins in hemocytes of cockroach *Leucophea maderae* L. *Dev. Comp. Immunol.* 2: 735-740.

Bellah, M.El.M., R.El Ridi, R. Abou-Elela and E.L. Cooper. 1988. Age-related occurrence of natural agglutinins in the Erisilk worm, *Philosamia ricini. Dev. Comp. Immunol.* 12: 707-717.

Bernheimer, A.W. 1952. Hemagglutinins in caterpillar blood. *Science* 115: 150-151.

Bradley, R.S., S.S. Gwendy, B. Stiles and K.D. Hapner. 1989. Grasshopper haemagglutinin: Immunochemical localization in haemocytes and investigation of opsonic properties. *J. Insect Physiol.* 35: 353-361.

Cohen, E. 1970. A review of the nature and significance of hemagglutinins of selected invertebrates. In: *Protein Metabolism and Biological Function*, p. 87. (C.P. Bianchi and R. Hilf, eds.). Rutgers Univ. Press, New Brunswick.

Denburg, J.L. 1980. Cockroach muscle haemagglutinins: Candidate recognition macromolecular. *Biochem. Biophys. Res. Com.* 97: 33-40.

Donlon, W.C. and C.T. Wemyss. 1976. Analysis of the haemagglutinin and general protein element of the haemolymph of the West Indian leaf cockroach, *Blaberus craniifer. J. Invert. Path.* 28: 191-194.

Drif, L. and M. Brehelin. 1989. Agglutinin mediated immune recognition in *Locusta migratoria* (Insecta). *J. Insect Physiol.* 35: 729-736.

Feir, D. and M.A. Walz. 1964. An agglutinating factor in insect haemolymph. *Am. Ent. Soc. Am.* 57: 388.

Gold, E.R. and P. Balding. 1975. Receptor-specific proteins, plant and animal lectins. *Excerpta Medica,* pp. 251-283. Elsevier, New York.

Goldstein, I.J., R.C. Huges, M. Monsingny, T. Osawa and N. Sharma. 1980. What should be called a lectin? *Nature* 285: 295.

Gupta, A.P. 1986. Arthropod immunocytes. In: *Hemocytic and Humoral Immunity in Arthropods,* pp. 13-59 (A.P. Gupta, ed.). John Wiley and Sons, New York.

Hapner, K.D. 1983. Hemagglutinin activity in the hemolymph of individual acrididae specimens. *J. Insect Physiol.* 29: 101-106.

Hapner, K.D. and M.A. Jermyn. 1981. Haemagglutinin activity in the haemolymph of *Teleogryllus commodus* (Walker). *Insect Biochem.* 11: 287-295.

Hapner, K.D. and M.R. Stebbins. 1986. Biochemistry of Arthropod agglutinins. In: *Hemocytic and Humoral Immunity in Arthropods,* pp. 227-250. (A.P. Gupta, ed.). John Wiley and Sons, New York.

Jurenka, R., K. Manfredi and K.D. Hapner. 1982. Haemagglutinin activity in Acrididae hemolymph. *J. Insect Physiol.* 28: 177-182.

Komano, H., D. Mizuno and S. Natori. 1980. Purification of lectin induced in the hemolymph of *Sarcophaga peregrina* larvae on injury. *J. Biol. Chem.* 255: 2919-2924.

Komano, H., D. Mizuno and S. Natori. 1981. A possible mechanism of induction of insect lectin. *J. Biol. Chem.* 256: 7087-7089.

Komano, H., R. Nozawa, D. Mizuno and S. Natori. 1983. Measurement of *Sarcophaga peregrina* lectin under various physiological conditions by radioimmunoassay. *J. Biol. Chem.* 258: 2143-2147.

Kubo, T., Komano, H., Okada, M. and S. Natori. 1984. Identification of haemagglutinating protein and bactericidal activity in the haemolymph of adult *Sarcophaga peregrina* on injury of the body wall. *Dev. Comp. Immunol.* 8: 283-291.

Kubo, T. and S. Natori. 1987. Purification and some properties of a lectin from the haemolymph of *Periplaneta americana* (American cockroach). *Eur. J. Biochem.* 168: 75-82.

Lackie, A.M. 1981. The specificity of the serum agglutinins of *Periplaneta americana* and *Schistocerca gregaria* and its relationship to the insect's immune response. *J. Insect Physiol.* 27: 139-143.

Lackie, A.M. and G.R. Vasta. 1988. The role of galactosyl-binding lectin in the cellular immune response of the cockroach *Periplaneta americana* (Dictyoptera). *Immunology* 64: 353-357.

Mauchamp, B. 1982. Purification of an N-acetyl-D-glucosamine specific lectin (P.B.A.) from epidermal cell membranes of *Pieris brassicae* L. *Biochimie* 64: 1001-1008.

McKenzie, A.N.J. and T.M. Preston. 1992a. Purification and characterization of a galactose specific agglutinin from the Haemolymph of the larval stages of the insect *Calliphora vomitoria, Develop. Comp. Immunol.* 16: 31-39.

McKenzie, A.N.J. and T.M. Preston. 1992b. Biological characteristics of the *Calliphora vomitoria* agglutinin. *Develop. Comp. Immunol* 16: 000-000 (in press).

Minnick, M.F., R.A. Rupp and K.D. Spence. 1986. A bacterial induced lectin which triggers hemocyte coaggulation in *Manduca sexta. Biochem. Biophys. Res. Commun.* 137: 729-735.

Noguchi, H. 1903. A study of immunization—haemolysins, agglutinins, precipitins and coagulins in cold-blooded animals. *Centralbl. f. Bakt. Abt. Orig.* 35: 352.

Olafsen, J.A. 1986. Invertebrate lectins: Biochemical heterogenecity as a possible key to their biological function. In: *Immunity in Invertebrates*, pp. 94-111. (M. Brehelin, ed.). Springer-Verlag, Berlin.

Pathak, J.P.N. 1991. Occurrence of natural hemagglutinins in *Halys dentata* (Hemiptera) during development. *Dev. Comp. Immunol.* 15: 99-104.

Pendland, J.C. and D.G. Boucias. 1985. Hemagglutinin activity in the hemolymph of *Anticarsia gemmatalis* larvae infected with the fungus *Nomuraea rileyi*. *Dev. Comp. Immun.* 9: 21-30.

Pendland, J.C. and D.G. Boucias, 1988. Function of galactose-binding lectin from *Spodoptera exigus* larval haemolymph: Opsonization of blastospores from entomogenous hyphomycetes. *J. Insect. Physiol.* 34: 533-540.

Pereira, M.E.A., A.F.B. Andrade and J.M.C. Ribeiro. 1981. Lectins of distinct specificity in *Rhodnius prolixus* interact selectively with *Trypanosoma cruzi*. *Science* 211: 597-599.

Ratcliffe, N.A. and A.F. Rowley. 1983. Recognition factors in insect hemolymph. *Dev. Comp. Immunol.* 7: 653-658.

Ratner, S. and S.B. Vinson. 1983. Phagocytosis and encapsulation: Cellular immune responses in Arthropoda. *Am. Zool.* 23: 185-194.

Renwrantz, L. and A. Stahmer. 1983. Opsonizing properties of an isolated haemolymph agglutinin and demonstration of lectin-like recognition molecule at the surface of haemocytes of *Mytilus edulis*. *J. Comp. Physiol.* 149: 535-546.

Renwrantz, L. and W. Mohr. 1978. Opsonizing effect of serum and albumin gland extract on the elimination of human erythrocytes from the circulation of *Helix pomatia*. *J. Comp. Physiol.* 141: 477-480.

Renwrantz, L., W. Schancke, H. Erl, H. Harm, H. Leibsch and J. Gércken. 1981. Descriminative ability and function of the immunobiological recognition system of the snail *Helix pomatia*. *J. Comp. Physiol.* 141: 477-488.

Scott, M.T. 1971. A naturally occurring agglutinin in the haemolymph of *Periplaneta americana*. *Arch. Zool. Exp. Gen.* 112: 73-80.

Sharon, N. 1984. Carbohydrates as recognition determinate in phagocytosis and in lectin-mediated killing of target cells. *Biol. Cell.* 51: 239-246.

Stebbins, M.R. and K.D. Hapner. 1985. Preparation and properties of haemagglutinin from haemolymph of Acrididae (grasshoppers). *Insect Biochem.* 15: 451-452.

Stebbins, M.R. and K.D. Hapner. 1986. Isolation, characterization and inhibition of arthropod agglutinins. In: *Hemocytic and Humoral Immunity in Arthropods*, pp. 463-491 (A.P. Gupta, ed.). John Wiley and Sons, New York.

Stein, E.A., A. Wojdani and E.L. Cooper. 1982. Agglutinins in the earthworm *Lumbrious terrestrie*: Naturally occurring and induced. *Dev. Comp. Immunol.* 6: 407-421.

Stiles, B., R.S. Bradley, G.S. Stuart and K.D. Hapner. 1988. Site of synthesis of the haemolymph agglutinin of *Melanoplus differentialis* (Acrididae: Orthoptera). *J. Insect Physiol.* 12: 1077-1085.

Stynen, D., M. Peferson and A. Deloof. 1982. Proteins with haemagglutinin activity in larvae of the Colorado beetle. *Leptinotarsa decemlineata*. *J. Insect Physiol.* 28: 165-470.

Suzuki, T. and S. Natori. 1983. Identification of a protein having hemagglutinating activity in the hemolymph of the silkworm, *Bombyx mori*. *J. Biochem.* 93: 583-590.

Takahashi, H., H. Komano, N. Kawaguchi, M. Obinata and S. Natori. 1984. Activation of the secretion of specific proteins from the fat body following injury to the body wall of *Sarcophaga peregrina* larvae. *Insect Biochem.* 14: 713-717.

Takahashi, H., H. Komano, N. Kawaguchi, N. Kitamura, S. Nakanishi and S. Natori. 1985. Cloning and sequencing of cDNA of *Sarcophaga peregrina* humoral lectin induced on injury of the body wall. *J. Biol. Chem.* 260: 12228-12233.

Takahashi, H., H. Komano and S. Natori. 1986. Expression of the lectin gene in *Sarcophaga peregrina* during normal development and under conditions where the defense mechanism is activated. *J. Insect. Physiol.* 32: 771-780.

Umetsu, K., S. Kosaka and T. Suzuki. 1984. Purification and characterization of a lectin from beetle *Allomyrina dichotoma*. *J. Biol. Chem.* 95: 239-245.

Wigglesworth, V.B. 1973. The role of the epidermal cells in moulding the surface pattern of the cuticle in *Rhodnius* (Hemiptera). *J. Cell Sci.* 12: 683-705.

Yeaton, R.W. 1981. Invertebrate lectins: II. Diversity of specificity, biological synthesis, and function in recognition. *Dev. Comp. Immunol.* 5: 535-545.

CHAPTER 12

Suppression of the Insect Immune System by Parasitic Hymenoptera

S. Bradleigh Vinson

Introduction

The invertebrate immune system remains poorly characterized. Although some of the mechanisms are just beginning to be examined, it is nonetheless clear that the vertebrate antigen-antibody system is not operational in invertebrates. Among the invertebrates, the class Insecta has received the greatest attention. But, in spite of extensive research, the immune system of insects is still poorly understood. It is not the purpose of this chapter to review the invertebrate or the insect immune system, as excellent reviews exist (Lackie 1988; Dunn 1986) and several edited books are available on the subject (Brehelin, 1986; Gupta, 1986), but rather to examine how parasitoids suppress the immune system of their insect host. As Bayne (1984) stated, understanding how parasites (parasitoids) can deal with the immune system of their host provides information on how the immune system functions. Thus what is known about the ability of parasitoids to suppress the insect immune system and some methods of study are examined here. However, before these topics are taken up, some background on the response of the insect immune system to parasitoids is required.

Immune Response to Parasitoids

Most internal parasitoids deposit their eggs within the haemocoel of their host (Sweetman 1958; Salt 1963). Within this environment are the nutritive supplies and environmental conditions necessary for the parasitoid's development. Nevertheless, the parasitoid must face three challenges in order to survive. The first is to initially successfully compete with and then to parasitize the host tissues for resources. The second is to process the available nutritional resources in such a way as to preserve the integrity of the environment (containers) until such time as the parasitoid can become self-sufficient. The third challenge is to deal effectively with the internal defensive system of the host. The immune system involves both humoral and cellular components. The humoral system generally refers to the biochemicals within the haemolymph of the organism that exert some effect on the invader or are involved in its recognition. These biochemicals include the inducible antibacterial factors (Boman *et al.*, 1986; Chadwick and Dumphy, 1986), lysozyme (Mohrig and Messner 1968), lectins (Takahashi *et al.*, 1986; Olafsen, 1986), acid mucopolysaccharides (Anderson and Chain, 1986) and the phenoloxidase-melanin system (Söderhäll and Smith, 1986). Of these biochemicals, the phenoloxidase system has been implicated most often in the immune reaction against mesozoan parasites (Brewer and Vinson, 1971; Dularay and Lackie, 1985).

The cellular components involve a number of single free-floating cells in the haemolymph collectively referred to as haemocytes. Considerable confusion exists as to the types of haemocytes present in the haemolymph. This confusion stems from differences in haemocyte morphology among different insect species, differences in the morphology of haemocytes from different ages of the same species, effects of different methods of study on haemocyte morphology and the effects of environmental and physiological factors on the morphology of the haemocytes. Such confusion has inhibited research on the characterization and function of the various haemocytes in insect immunity, although the situation is changing and some theories have been formulated. In most cases the cellular response to invading mesozoan parasites is a process of encapsulation (Salt, 1963). The process of encapsulation is unique among the arthropods (Baerwald, 1979), arising from the joint participation of cells that normally exist as monodispersed cells in the haemolymph of the insect. Encapsulation is believed to be similar to phagocytosis wherein foreign material is recognized, attaches to the cell membrane, is engulfed and subsequently slowly digested (Gupta, 1985). However, during encapsulation the object is too large to be engulfed; thus the haemocytes spread out over the surface, resulting in the foreign object becoming totally covered by a layer of haemocytes. Usually the capsules constitute several layers of adhering haemocytes with the outermost haemocytes being less deformed in the process of attachment to the capsule and some returning to circulation (Götz, 1986). During the last decade, the theory of encapsulation has evolved (Brehelin *et al.*, 1975; Götz, 1986; Gupta, 1985; Ratcliffe and Rowley, 1979; Schmidt and Ratcliffe, 1977) as follows: (a) the foreign material is either randomly contacted by granulocytes or the granulocytes are attracted to the object, which is recognized as foreign, resulting in granulocyte degranulation; (b) the released material interacts with the foreign, object surface making it sticky and attracting more granulocytes and plasmatocytes; (c) the plasmatocytes then attach to the foreign object, aided by the granulocyte released factor, coating the foreign object; (d) the plasmatocytes, once attached, spread over the surface, cutting off oxygen and nutrition (Götz and Boman, 1985; Gupta, 1985). In many cases the capsule melaninizes (Götz and Boman, 1985) but in some host species a melanotic capsule develops that is not composed of cells (Götz *et al.*, 1977; Götz and Boman, 1985).

Circumventing the Host Defence System

Insects possess a number of means for inhibiting the success of insect parasitoids. These include both external and internal physiological factors (Vinson, 1990) as well as behavioural; the internal factors are the last defence against invading parasitoids. Salt (1968) divided the methods of immune resistance by insects to parasitoids into six categories. His research and insight provided an excellent foundation for our present understanding of how parasitoids interact with the cellular insect immune system. However, his categories overlap and provide little information as to the strategies evolved by the parasitoids to circumvent the system. Vinson (1990) categorized the different means devised for circumventing the immune systems of their host into five categories: (1) avoidance, (2) evasion, (3) destruction, (4) suppression, and (5) subversion. Avoidance is accomplished by those species which oviposit their eggs in the eggs or tissues of the host, which do not mount a classical immune defence, or which are ectoparasitoids. A few species, particularly larval dipterans, subvert the system so that a haemolytic capsule forms, but the parasitoid develops within the capsule while being maintained by an outside air supply. The remainder of the endoparasitic species must either evade, suppress or destroy the immune system.

Parasitoids may evade the immune system in two ways. One is to evolve or acquire surface characteristics that are identical to or very similar to the surfaces within the host so that the latter will fail to respond to the foreign surface because it appears to be 'self'. This method of evasion has been referred to as molecular mimicry (Damian, 1964). Another approach is to evolve a surface that may differ but nevertheless fail to elicit a response from the host. This has been referred to as cloaking (stealth), whereby the foreign surface remains unrecognized and thus hidden from the immune system (Vinson, 1990). Although it is still difficult to distinguish between these two approaches without an understanding of the molecular aspects, there is increasing evidence that some parasitoids or at least their egg stage evade the insect immune system.

The eggs of such parasitoids as *Telenomus heliothidis* (Scelionidae) and *Trichogramma pretosum* (Chacididae) have a thin chorion while species such as *Cardiochiles nigriceps* (Braconidae) or *Venturia (= Nemeritus) canescens* (Ichneumonidae) have a thicker chorion, consisting of a fibrous outer coat

(King *et al.*, 1969; Rotheram, 1973a; Vinson and Scott, 1974a). One difference between the former and latter species, besides the nature of the chorion, is that the latter two species deposit their eggs in the haemocoel, thus exposing them to the host's immune system. There is evidence that the fibrous outer coat is not recognized as foreign and as long as the fibrous layer remains, the eggs are not encapsulated (Brewer *et al.*, 1972; Davies and Vinson, 1986; Davies *et al.*, 1986). Rotheram (1973a and b), working with *V. canescens*, suggested that small membrane-bound particles (referred to as virus-like particles by Schmidt *et al.*, 1990) occur embedded in the outer fibrous layer of the egg chorion and block or suppress the host's immune system. These virus-like particles are approximately, 1,300°A across and simultaneously injected into the host along with the egg (Salt, 1965). More recent studies with the virus-like particles (Feddersen *et al.*, 1986; Schmidt *et al.*, 1990; Schmidt and Schuchmann-Feddersen, 1989) have demonstrated that these particles from the calyx of *V. canescens* consist of several proteins, one of which appears to be similar to the proteins occurring in the basal lamina of the host. These results suggest that this parasitoid might evade the host's immune system by molecular mimicry. However, Berg *et al.* (1988) found that the titre of the basal lamina protein increased upon exposure to the particles from *V. canescens*. These results indicate that the invasion was recognized because the host's immune system reacted; thus the particles might interfere with either the recognition system or the ability of the host to respond.

Host Immune Suppression

There is increasing evidence that parasitoids inject factors that suppress the immune response in their host. Three different systems have been studied to date (Vinson, 1990). The first involves the braconid pupal parasitoid *Pimpla turionellae*, which successfully attacks the pupae of various lepidopterans (Krombein *et al.*, 1986). In a series of papers (Osman, 1974, 1978; Osman and Führer, 1979), the plasmatocytes were found to be inhibited in their ability to form pseudopodia when parasitized. Three glands from the female parasitoid were identified as involved: the uterus (oviduct) gland, containing an acid-mucopolysaccharide and a lipoprotein; the poison gland, containing a neutral mucopolysaccharide; and Dufour's gland, containing a cholesterol ester and a lecithin-like phospholipid. How these compounds act and whether all or just some of the compounds are needed is not known.

The second system involves the cynipid parasitoid *Leptophilina heterotoma* (=*Pseudeucoila bochei*) and its *Drosophila* larval host. In *Drosophila* there are basically two types of haemocytes, the crystal cells and plasmatocytes (Rizki, 1957). During development the plasmatocytes gradually transform through a stage, referred to as a podocyte, to form a large, flat, disc-shaped cell referred to as a lamellocyte (Rizki, 1957;

Nappi, 1973). By pupation most of the plasmatocytes have become lamellocytes and the crystal cells have disappeared (Rizki, 1957; Nappi and Streams, 1969). When *Drosophila* larvae were attacked by a permissive parasitoid, such as *L. heterotoma*, an increase in crystal cells and a decrease in lamellocytes occurred, suggesting the parasitoid had suppressed the immune response (Nappi and Streams, 1969). However, the effect was more dramatic when compared to the haemocyte changes that occurred in resistant host, i.e., hosts that mounted an effective defence reaction. In these hosts there was a marked increase in the total haemocyte population, a decrease in crystal cells and a precocious increase in lamellocytes (Walker, 1959; Nappi and Streams, 1969). It was found that the female parasitoid injects a substance (Streams and Greenberg, 1969) that inhibits lamellocyte adhesiveness and reduces lamellocyte differentiation (Rizki and Rizki, 1984; Walker, 1959). Factors, apparently from the poison gland, affected microtubule formation (Rizki and Rizki, 1990a) and appeared to confer cross-protection to other *Drosophila* parasitoids (Nappi, 1974; Rizki and Rizki, 1986; Streams and Greenberg, 1969). It is particularly interesting that virus-like particles have been reported from the accessory gland of *L. heterotoma* that infect lamellocytes but not plasmatocytes (Rizki and Rizki, 1990b). These authors suggest that the virus-like particles may determine the fate of infected host blood cells. As discussed below, the results suggest that a common approach to immune suppression by several parasitic hymenoptera has evolved.

The third system occurs in two parasitoid families, the endoparasitic ichneumonids and the endoparasitic braconids. Within some genera of these families, occurs what has been described as a symbiotic relationship between a virus and the parasitoid (Vinson and Stoltz, 1986). In these parasitoid species the virus exists as an enveloped one, consisting of segmented, double-stranded, circular DNA genomes. This virus has been established as a new family of insect virus, the Polydnaviridae (Stoltz *et al.*, 1984). Polydnaviruses have only been isolated from the calyx or lateral oviducts (Stoltz and Vinson, 1979b), the virus being assembled in the calyx epithelium. This epithelium occurs just below the egg tube pedicle and comprises the upper epithelium of the calyx. In braconids the calyx comprises the upper lateral oviduct and is variously expanded, as shown in the example from *Cardiochiles abdominalis* (Fig. 1). In the ichneumonids the calyx and the remainder of the lateral oviduct are similar, with the calyx epithelium comprising the distal lateral oviduct epithelium (Fig. 2). The virus, once assembled, migrates from the nucleus of the epithelial cell through the cytoplasm and is released in the lumina (Norton *et al.*, 1975; Norton and Vinson, 1983). One to several virions, depending on the species, exist within one unit membrane in the polydnavirus from the braconids and within two unit membranes in the polydnavirus from the ichneumonids (Stoltz and Vinson, 1979b). The virions are so numerous

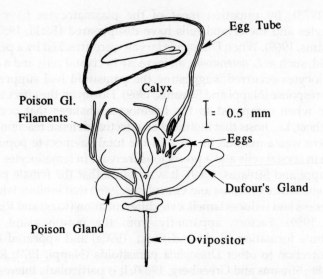

Fig. 1: The reproduction system of *Cardiochiles abdominalis* is representative of many braconids showing the Dufour's gland, poison gland with associated filaments and ovary consisting of the egg tube, expanded calyx (found in some species), and the lateral and common oviduct (not labelled) leading to the ovipositor.

within the calyx of most species that they constitute a particulate fluid (calyx fluid), which resembles a thick paste upon rupture of the calyx (Fig. 2). The virus has been injected into host insects using a glass micropipette mounted on a micromanipulator. (Stoltz and Vinson, 1979a and b). The virus is expressed (Fleming *et al.*, 1983; Blissard *et al.*, 1986) but does not replicate (Theilmann and Summers, 1986) in the host. The virus has many effects (Stoltz, 1986a; Vinson and Dahlman, 1989) including effects on the immune system of the host (Edson *et al.*, 1981).

The role of the polydnaviruses in suppressing the insect's immune system evolved out of research initiated by Dr. G. Salt. Salt (1973) reported the presence of a factor in the calyx of a parasitoid that imparted resistance to the host's immune response on the part of the egg. The factor from *V. canescens* was found to be produced in the nucleus of the calyx epithelium and was reported to consist of proteins (Rotheram, 1973a and b; Bedwin, 1979). A factor was also isolated from the calyx of *C. nigriceps* and later *Campoletis sonorensis* that was produced (Vinson and Scott, 1974a; Norton *et al.*, 1975) and acted similarly (Vinson, 1972) to that reported for *Venturia* by Salt. However, the particles from the calyx of *C. nigriceps* were found to contain DNA (Vinson and Scott, 1974b). The virus-like particles of *V. canescens* were later reported to also contain small amounts of DNA (Federici and Suduth-Klinger 1987), but these results could not be substantiated (Schmidt and Schuchmann-Feddersen, 1989). Similar DNA

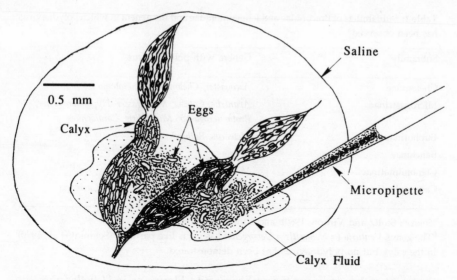

Fig. 2: The reproductive system of a representative ichneumonid placed in saline on a glass slide and into which the calyx fluid and eggs flow from the ruptured lateral oviducts. The calyx fluid can be collected in a micropipette.

viruses, however, have also been reported from the egg-larval parasitoid *Phanerotoma flavitestacea* (Poinar *et al.*, 1976; Hess *et al.*, 1979). Since these early studies, polydnaviruses have been reported from all species within several subfamilies of the ichneumonids with the exception of the genus *Venturia* (Table 1). In addition to the polydnaviruses found in a number of genera of ichneumonids and braconids (Stoltz, 1986b; Stoltz and Vinson, 1979b), an entomopoxvirus has been reported from the poison glands of braconids of the subfamily Opininie (Edson, 1981; Lawrence and Akin, 1990). Although the role of the polydnaviruses in suppressing the immune system of the hosts in which they were injected has been demonstrated by Edson *et al.* (1981), the role of the entomopoxvirus in the host-parasitoid relationship or whether it affects the host's immune system remains speculative.

Several studies on the mechanism of polydnavirus suppression of the immune system (Davies *et al.*, 1987; Davies and Vinson, 1988; Tanaka, 1987b; Stoltz and Guzo, 1986) have demonstrated that exposure of the host to a polydnavirus results in plasmatocyte abnormalities (reduced spreading). Davies and Vinson (1988) presented evidence that the reduced haemocyte spreading was due to a factor in the haemolymph and did not appear to be due to direct viral exposure. Although Guzo and Stoltz (1987) found that direct exposure of cells in an insect tissue culture to the polydnavirus of *Cotesia melanoscela* did induce morphological changes, the significance of this result was difficult to assess. They also suggested

Table 1: Subfamilies of Braconidae and Ichneumonidae and the genera in which polydnavirus has been observed[a]

Subfamily	Genera with polydnavirus
Cheloninae	*Ascogaster, Chelonus, Phanerotoma*
Microgastrinae	*Apantelis, Cotesia, Microgaster, Hypomicrogaster, Protomicrogaster, Microplitis, Cardiochiles*
Euphorinae	*Meteorus, Bathyplectes*
Banchinae	*Glypta*
Ctenopilmatinae	*Misoleius*
Campopleginae[b]	*Diadegma, Campoletis, Hyposoter, Casinaria*

[a]*Source* : Stoltz and Vinson, 1969b and Vinson, unpubl.
[b]The genus *Venturia* (= *Nemeritis* = *Devorgella*) has virus-like particles (Schmidt *et al.*, 1990) in the calyx but no polydnavirus has been demonstrated.

the target of the virus was the prohaemocyte. However, in *Heliothis virescens* the polydnavirus of *C. sonorensis* (CsV) did not appear to affect the prohaemocytes, but resulted in both a decrease in the number of plasmatocytes and a reduction in the ability of the remaining plasmatocytes to spread (Davies and Vinson, 1988). In *Spodoptera frugiperda* the development of the parasitoid *C. sonorensis* was only partially successful with approximately 30% of the parasitized larvae failing to produce parasitoids. The CsV only inhibited the encapsulation of the parasitoid eggs in 68% of the treatments, which was accompanied by reduced ability of the haemocytes to spread. In the remaining hosts (32%) no reduction in haemocyte spreading ability was demonstrated and in such cases the parasitoid eggs were encapsulated (Provost *et al.*, 1990). These results support the hypothesis that reduced haemocyte spreading is effective in suppressing the ability of the host immune system to encapsulate foreign material. They also explain the results of earlier research in which one species could protect another from encapsulation (Vinson, 1977; Vinson and Stoltz, 1986; Streams and Greenberg, 1969).

How the polydnavirus inhibits plasmatocyte spreading has yet to be elucidated. Davies and Vinson (1988) provided evidence that a factor in the haemolymph, rather than the polydnavirus per se was involved in inhibiting plasmatocyte spreading, and that the latter could be restored with haemolymph from healthy hosts. It is possible that the CsV inhibits microtubule formation, thus acting similarly to lamellolysin (Rizki and Rizki, 1986), which suppresses the immune response of *Drosophila*. Preliminary studies of the CsV induced factor indicated the haemolymph factor was heat sensitive and destroyed by trypsin (Vinson and Davies, unpubl.).

Several studies have shown that 'new' proteins occur in the haemolymph of parasitized hosts soon after parasitism (Cook *et al.*, 1984; Beckage *et al.*, 1989). However, there is no clear evidence that these proteins are involved in immune suppression.

There are also other factors from other parasitoid sources that have been implicated in suppressing the host's immune response. Vinson (1972) reported that the encapsulation of *C. nigriceps* larvae injected into hosts priorly injected with teratocytes was reduced. Salt (1971) had earlier suggested that teratocytes released in a host inhibit its immune response through attrition of biochemicals needed to elicit such a response. But how the specificity of the immune inhibition could be accounted for by such attrition is not known. Tanaka (1987a and c) reported that both teratocytes and the calyx fluid were necessary to protect the eggs and larvae of *Microplitis mediator* from encapsulation. For *Apanteles kariyai* the teratocytes are implicated in suppressing the host's immune system along with both the calyx fluid and venom (Tanaka, 1987a; Tanaka and Wago, 1990). Kitano (1986) also believed that the venom of *A. glomeratus*, together with the calyx fluid, was involved in suppressing the immune system but suggested that the venom acted at the egg's surface. However, in the braconids the venom might be necessary to enable the virus to penetiate the host cells (Stoltz *et al.*, 1988), which would explain the synergistic effects of virus and venom in a number of the braconid systems studied (Kitano, 1982; Guzo and Stoltz, 1987; Tanaka, 1987c). This, however, does not explain the role of the teratocytes.

The polydnavirus appears to be integrated into the genome of the parasitoid since it is found in both males and females (Flemming and Summers, 1988). These data suggest that viral sequences are probably present in all parasitoid derived cells, including teratocytes. It has been suggested that both the venom gland (Webb and Summers, 1990) and teratocytes (Dahlman, 1990) may release viral products. Certainly, if the viral genome exists as an integrated sequence in the parasitoid's genome, the potential for any parasitoid tissue to duplicate the role of the virus must also be acknowledged.

Methods of Isolating the Polydnavirus for Study

The role of the polydnavirus in suppressing the immune system of its host has only been examined in a few parasitoid species. Whether the polydnavirus from other parasitoid species would have a similar effect on the immune system of other hosts is not known. Such information is hardly trivial since the proteins and circular DNA of the polydnaviruses from different species could well differ (Stoltz, 1986b). In order to determine the effects of the polydnaviruses on the immune or other physiological systems of the host, the virus must be isolated. This can be accomplished by first removing the reproductive system of the female. We used middle-aged females, held without ovipositing and rinsed in 2% NaClO to reduce contamination. Using forceps to tear the intersegmental membranes between the posterior abdominal and remaining anterior segments (Fig. 3A) aided in pulling the reproductive system free of the abdomen. Pinning the female down in a wax dish or vising her in a forceps, the ovipositor can be grasped by another forceps (Figure 3B) and pulled free to remove the reproductive system. The procedure usually removes the ovaries, Dufour's gland and poison gland (Fig. 1).

Once removed, the ovaries should be rinsed in cold Pringle's saline (Pringle, 1938) or in phosphate buffered saline (PBS). If crude polydnavirus is desired, which may contain several types of virus, various proteins and other biochemicals, the ovaries should be placed on a microscope slide in cold saline (approximately an equal volume). The presence of a cloudy to bluish tinted fluid around the egg or filling the lumina of the calyx or lateral oviduct is an indication that a virus is present. The epithelium is opened, allowing the paste (fluid) and eggs, if any, to flow into the surrounding saline (Fig. 2). The ovary tissue is then removed and the virus-containing solution collected with a micropipette and further purified or used in various studies as a crude calyx fluid. Alternately, if larger amounts are desired, we place the ovaries in 1.5 ml microcentrifuge tubes (20 + ovaries) with a minimal amount of saline (1-2 µl per pair of ovaries). The ovaries are macerated with microdissection scissors and the material is centrifuged at 50 g for 2 minutes to remove debris. The result is crude calyx fluid. The amount of virus can be standardized by placing the calyx fluid in a 1 µl capillary pipette, 1/4 full, sealing it and placing it in a glass sleeve for support. The tube is centrifuged at 45 °C in a fixed angle rotor microcentrifuge at 8000 g for 2 minutes and then immediately centrifuged in a swinging bucket rotor for 5 minutes at 975 g to form a perpendicular

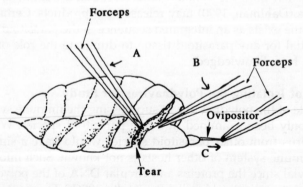

Fig. 3: The abdomen of an adult parasitoid showing a method to remove the reproductive system. The last and next to last abdominal ventral sclerites are grasped with forceps (A&B) and pulled apart to tear the segments apart, then the ovipositor is pulled out using forceps "B" applied at "C".

interface. The length of the pellet can be measured and related to the length of the pellet plus supernatant to obtain the concentration of viral particles (Dover *et al.*, 1988), which can be further diluted.

If purified polydnavirus is to be used, the ovaries or calyx region can be placed in small Eppendorf microphage tubes for 24 hours in 1-2 ml phosphate buffer saline per ovary. Depending on the species, from 50 to several hundred ovaries may be required to obtain the requisite amount of virus. Once enough ovaries have been collected, they are lightly macerated as described for the alternate procedure above. The supernatant can be further purified by centrifugation at 8000 g for 2 minutes, the pellet rinsed in buffer and resuspended in buffer. The result is purified calyx fluid. The supernatant (containing the virus) can be further purified through density gradient separation. A sucrose gradient (wt/wt) from 25 to 50% in PBS is made and the viral supernatant is carefully layered on top and centrifuged at 125,000 g (in a swinging bucket rotor) for 60 minutes at 5°C. The resultant virus band is collected and dialyzed against PBS for 24 hours. This purified virus can be used to determine the effects of the virus on the immune or other host systems.

The virus can be injected into hosts via micropipette injectors. These can be made by pulling a 1.5 mm OD ≃ 0.55 mm ID) capillary tube into a flame to a diameter of about 0.2 mm ID. These tubes (About 2 cm long) are glued into 2 cm long sections of gum rubber tubing. We have found that flexible silicone glue is effective in cementing the capillary tube into one end of the gum rubber tubing and sealing the other end. Once made, the micropipette injector can be calibrated (Fig. 4). Injections are best made on CO_2 narcotized insects to minimize the danger of bleeding. Injections into lepidopterans are often better accomplished if the injection site is just behind the head capsule with the injector tip placed parallel to and just

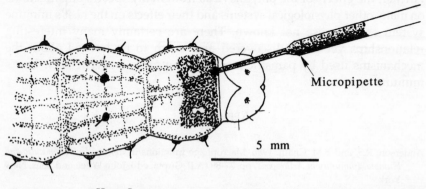

Micropipette

5 mm

Host Larva

Fig. 4: A narcotized host larvae being injected with calyx fluid via a micropipette inserted between the dorsal first thoracic sclerite and head capsule.

under the cuticle (Fig. 4). Once inserted, the microinjector is removed after 10 seconds and larvae allowed to feed. The virus enter in the host tissues in approximately six hours (Stoltz and Vinson, 1979a; Blissard *et al.*, 1986).

How parasitoids evade the immune system of their host may differ in detail from one species complex to another. However, some patterns have begun to emerge. In *C. nigriceps* the egg appears to be protected by a mucoprotein outer coat (Davies and Vinson, 1986). In a permissive host, the coat persists until the larva hatches from the egg. However, in a resistant host, the protective coat is lost and the egg is encapsulated (Vinson and Scott, 1974a). Even in permissive hosts the protective outer coat of the egg slowly disappears, which suggests a race between the host to overcome the invader and the developing larva to hatch. The egg protection also provides time for the polydnavirus to invade and express (Blissard *et al.*, 1986). Once the larva has emerged from the egg, it is liberated into an environment in which the ability of the plasmatocytes to spread has been abrogated (Davies and Vinson, 1988). The evidence suggests that the ability of the plasmatocytes to spread is inhibited by a factor in the haemolymph (Davies and Vinson, 1988).

Suppression of the ability of the plasmatocyte-type cells of the host to spread is a common situation among several of the few subfamilies of parasitoids studied, for example, the cynipid, *Leptopilina* (Rizki and Rizki, 1984), and the braconid, *Pimpla* (Osman and Führer, 1979). Whether similar factors are involved in suppressing the haemocytes or some other factors which affect a similar target is not yet known.

Conclusion

The role of the polydnavirus in immune suppression is not yet clear. Further, the effects of the polydnavirus from many species of parasitoids on many other physiological systems and their effects on the host's immune system are likewise not known. There are certainly many interesting relationships yet to be discovered and much to be learned about the mechanisms used by parasitic insects to suppress or evade their host's immune system.

REFERENCES

Anderson, R.S. and B.M. Chain. 1986. Macrophage functions in insects. In: *Haemocytic and Humoral Immunity in Arthropods*, pp. 77-89 (A.P. Gupta, ed.). John Wiley and Sons, New York.

Baerwald, R.J. 1979. Fine structures of hemocyte membranes and the intercellular junctions formed during hemocyte encapsulation. In: *Insect Hemocytes*, pp. 231-258 (A.P. Gupta, ed.). Cambridge University Press, Cambridge.

Bayne, C.J. 1984. Immunoparasitology: Invertebrates. *Dev. Comp. Immunol.* (Suppl.) 3: 197-206.

Beckage, N.E., D.J. Nesbit, B.D. Nielsen, K.D. Spence and M.A.E. Barman. 1989. Alternation of hemolymph polypeptides in *Manduca sexta* larvae parasitized by *Cotesia congregata*: A two-dimensional electrophoretic analysis and comparison with major bacteria-induced proteins. *Arch. Insect Biochem. Physiol.* 10: 29-45.

Bedwin, O. 1979. An insect glycoprotein: A study of the particles responsible for the resistance of the parasitoid's egg to the defense reactions of its insect host. *Proc. Roy. Soc. Lond. (Biol.)* 205: 271-286.

Berg, R., I. Schuchmann-Feddersen and O. Schmidt. 1988. Bacterial infection induces a moth protein which has antigenic similarity to virus-like particle protein of a parasitoid wasp. *J. Insect. Physiol.* 34: 473-480.

Blissard, G.W., J.G.W. Fleming, S.B. Vinson and M.D. Summers. 1986. *Campoletis sonorensis* virus: Expression in *Heliothis virescens* and identification of expressed sequences. *J. Insect Physiol.* 32: 351-359.

Boman, H.G., I. Faye, P.V. Hofsten, J.Y. Lee, K.G. Xanthopoulos, H. Bennich, A. Engstrom, B.R. Merrifield and D. Andrew. 1986. Antibacterial immune proteins in insects. A review of some current perspectives. In: *Immunity in Invertebrates: Cells, Molecules and Reactions*, pp. 63-73 (M. Brehelin, ed.). Springer-Verlag, Berlin.

Brehelin, M., J.A. Hoffmann, G. Matz and A. Porte. 1975. Encapsulation of implanted foreign bodies by hemocytes in *Locusta migratoria* and *Melolontha melolontha*. *Cell Tissue Res.* 160: 283-289.

Brehelin, M. (ed.) 1986. Immunity in Invertebrates. Springer-Verlag, Berlin, Heidelberg.

Brewer, F.D. and S.B. Vinson. 1971. Chemicals affecting the encapsulation of foreign material in an insect. *J. Invertebr. Pathol.* 18: 287-289.

Brewer, F.D., B. Glick and S.B. Vinson. 1972. Immunological investigations of the factors responsible for the resistance of *Heliothis zea* and susceptibility of *Heliothis virescens* to the parasitoid *Cardiochiles nigriceps*. *Comp. Biochem. Physiol.* 43: 781-786.

Chadwick, J.S. and G.B. Dumphy. 1986. Antibacterial and antiviral factors in arthropod hemolymph. In: *Haemocytic and Humoral Immunity in Arthropods*, pp. 287-330 (A.P. Gupta, ed.). John Wiley and Sons, New York.

Cook, D.I., D.B. Stoltz and S.B. Vinson. 1984. Induction of a new hemolymph glycoprotein in larvae of permissive host parasitized by *Campoletis sonorensis*. *J. Insect Biochem.* 14: 45-50.

Dahlman, D.L. 1990. Evaluation of teratocyte functions: An overview. *Arch Insect Biochem. Physiol.* 13: 159-166.

Damian, R.T. 1964. Molecular mimicry: Antigen sharing by parasite and host and its consequences. *Am. Naturalist* 98: 129-149.

Davies, D.H. and S.B. Vinson. 1986. Passive evasion by eggs of braconid parasitoid *Cardiochiles nigriceps* of encapsulation *in vitro* by haemocytes of host *Heliothis virescens*. Possible role for fibrous layer in immunity. *J. Insect. Physiol.* 32: 1003-1010.

Davies, D.H. and S.B. Vinson, 1988. Interfrence with functions of plasmatocytes of *Heliothis virescens in vivo* by calyx fluid of the parasitoid *Campoletis sonorensis*. *Cell Tissue Res.* 251: 467-475.

Davies, D.H., R.L. Burghardt and S.B. Vinson. 1986. Oogenesis of *Cardiochiles nigriceps* Vierick (Hymenoptera: Braconidae): Histochemistry and development of the chorion with special reference to the fibrous layer. *Int. J. Insect. Morphol. Embryol.* 15: 363-374.

Davies, D.H., M.R. Strand and S.B. Vinson. 1987. Changes in differential hemocyte count and *in vitro* of plasmatocytes from *Heliothis virescens* caused by *Campoletis sonorensis* polydnavirus. *J. Insect. Physiol.* 33: 143-154.

Dover, B.A., D.H. Davies and S.B. Vinson. 1988. Dose-dependent influence of *Campoletis sonorensis* polydnavirus on the development and ecdysteroid titers of last-instar *Heliothis virescens* larvae. *Arch. Insect. Biochem. Physiol.* 8: 113-126.

Dularay, B. and A.M. Lackie. 1985. Haemocytic encapsulation and the prophenoloxidase activation pathway in the locust, *Schistocerca gregaria* Forsh. *Insect Biochem.* 15: 827-834.

Dunn, P.E. 1986. Biochemical aspects of insect immunology. *Ann. Rev. Entomol.* 31: 321-339.

Edson, K.M. 1981. Virus-like and membrane-bound particles in the vemon apparatus of a parasitoid wasp. *Proc. 39th Ann. Electron. Microscop. Soc. Amer. Meeting*, p. 610.

Edson, K.M., S.B. Vinson, D.B. Stoltz and M.D. Summers. 1981. Virus in a parasitoid wasp: Suppression of the cellular immune response in the parasitoid's host. *Science* 211: 582-583.

Fedderson, I., K. Sander and O. Schmidt. 1986. Virus-like particles from host protein-like antigen determinants protect an insect parasitoid from encapsulation. *Experientia* 42: 1278-1281.

Federici, B.A. and J.M. Suduth-Kiinger. 1987. Evidence for DNA in the calyx particles of *Venturia canescens*. In: *Parasitoid Insects*, p. 101 (E. Bouletreau and G. Bonnot, eds.). Les Colloques de L'INRA. INRA Publ., Paris.

Fleming, J.G.W. and M.D. Summers. 1988. *Campoletis sonorensis* endoparasitic wasps contain integrated and extrachromosomal polydnavirus DNA. *J. Virol.* 57: 552-562.

Fleming J.G.W., G.W. Blissard, M.D. Summers and S.B. Vinson. 1983. Expression of *Campoletis sonorensis* virus in the parasitized host, *Heliothis virescens*. *J. Virol.* 48: 74-78.

Götz, P. 1986. Encapsulation in arthropods. In: *Immunity in Invertebrates*, pp. 153-170 (M. Brehelin, ed.). Springer-Verlag, Berlin.

Götz, P. and H. Boman. 1985. Insect immunity. In: *Comprehensive Insect Physiology, Biochemistry and Pharmacology*, vol. 3, pp. 453-485 (G.A. Kerkut and L.I. Gilbert, eds.). Pergamon Press, New York.

Götz, P., J. Roettgen and W. Lingg. 1977. Encapsulation humoral en tant que reaction de difense chez les Dipteres. *Ann. Parasitol. Hum. Comp.* 52: 95-97.

Gupta, A.P. 1985. Cellular elements in the hemolymph. In: *Comprehensive Insect Physiology Biochemistry, and Pharmacology*, vol. 3, pp. 401-451 (G.A. Kerkut and L.I. Gilbert, eds.). Pergamon Press, New York.

Gupta, A.P. (ed.). 1986. *Haemocytic and Humoral Immunity in Arthropods*. John Wiley and Sons, New York, 535 pp.

Guzo, D. and D.B. Stolz. 1987. Observations on cellular immunity and parasitism in the tussockmoth. *J. Insect Physiol.* 33: 19-31.

Hess, R.T., G.O. Poinar Jr. and L.E. Caltagirone. 1979. DNA-containing particle in the calyx of *Phanerotoma flavitestacea*. *J. Inertebr. Pathol.* 33: 129-132.

King, P.E., J.G. Richards and M.J.W. Copland, 1985. The structure of the chorion and its possible significance during oviposition in *Nasonia vitripennis* (Walker) and other chalcids. *Proc. R. Entomol. Soc. Lond.* (A)43: 13-20. ‵

King, P.E., N.A. Ratcliffe and M.J.W. Copland. 1969. The structure of the egg membranes in *Apanteles glomeratus* (L.). *Proc. R. Entomol. Soc. Lond.* (A)44: 137-142.

Kitano, H. 1982. Effect of the venom of the gregarious parasitoid, *Apanteles glomeratus* on its haemocytic encapsulation by the host, *Pieris rapal crucivora J. Invertebr. Pathol.* 40: 61-67.

Kitano, H. 1986. The role of *Apanteles glomeratus* venom in the defensive response of its host, *Pieris rapae crucivora*. *J. Insect Physiol.* 32: 369-375.

Krombein, K.V., P.D. Hurd Jr. and D.R. Smith. 1986. Catalog of Hymenoptera in America North of Mexico, vol. 3:2735.2 Smithsonian Institution Press, Washington, D.C.

Lackie, A.M. 1988. Mehocyte behaviour. *Adv. Insect Physiol.* 21: 85-178.

Lawrence, P.O. and D. Akin. 1990. Virus-like particles from the poison glands of the parasitic wasp. *Biosteres longicaudatus*. *Can. J. Zool.* 68: 539-546.

Mohrig, W. and B. Messner. 1968. Immureaktionen bei Insekten. II Lysozym als antimikrobielles Agents im Darmtrakt von Insekten. *Sondrrb. Biol. Zbl.* 87: 705-718.

Nappi, A.J. 1973. Haemocytic changes associated with encapsulation and melanization of some insect parasites. *Exptl. Parasitology*, 33: 285-302.

Nappi, A.J. 1974. Insect hemocytes and the problem of host recognition of foreignness. In: *Contemporary Topics in Immunobiology*, vol. 4, pp. 207-227 (E.L. Cooper, ed.). Plenum Press, New York.

Nappi, A.J. and F.A. Streams. 1969. Haemocytic reactions of *Drosophila melanogaster* to the parasites *Pseudeucoila mellyses* and *P. bochei*. *J. Insect Physiol.* 15: 1551-1566.

Norton, W.N. and S.B. Vinson. 1983. Correlating the initiation of virus replication with a specific pupal developmental phase of an ichneumonid parasitoid. *Cell Tissue Res.* 231: 387-389.

Norton, W.N., S.B. Vinson and D.B. Stoltz. 1975. Nuclear secretory particles associated with the calyx cells of the ichneumonid parasitoid *Campoletis sonorensis* (Cameron). *Cell Tissue Res.* 162: 195-208.

Olafsen, J.A. 1986. Invertebrate lectins—biochemical heterogeneity as a possible key to their biological function. In: *Immunity in Invertebrates: Cells, Molecular and Defense Reactions*, pp. 94-111 (M. Brehelin, ed.). Springer-Verlag, Berlin.

Osman, S.E. 1974. Parasitentoleranz von Schmetterlingspuppen Maskierung der Parasiteneier mit Mucopolysaccariden. *Naturwissenschaften* 61: 453-457.

Osman, S.E. 1978. Die Wirkung der Sekrete der weiblichen Genitalanhangsdrusen von *Pimpla turionellae* L. (Hym., Ichneumonidae) anf die Hamocyten und die Einkapslungsreaktion von Virtspuppen. *Z. fur Parasitenk.* 57: 89-100.

Osman, S.E. and E. Führer. 1979. Histochemical analysis of accessory genital gland secretions in female *Pimpla turionellae* L. (Hymenoptera: Ichneumonidae). *Int. J. Invertebr. Reprod.* 1: 323-332.

Poinar Jr., G.O., R. Hess and L.E. Caltagirone. 1976. Virus-like particles in the calyx of *Phanerotoma flavitestacea* and their transfer into host tissues. *Acta. Zool. Stockh.* 57: 161-167.

Pringle, J.W.S. 1938. Proprioception in insects. *J. Exp. Biol.* 15: 101-113.

Provost, G., D.H. Davies and S.B. Vinson. 1990. Evasion of encapsulation by parasitoid correlated with the extent of host hemocyte pathology. *Entomol. Exp. Appl.* 55: 1-10.

Ratcliffe, N.A. and A.F. Rowley. 1979. Role of hemocyte in defense against biological agents. In: *Insect Hemocytes*, pp. 331-414 (A.P. Gupta, ed.). Cambridge University Press, Cambridge.

Rizki, T.M. 1957. Alternation in the hemocyte population of *Drosophila melanogaster. J. Morph.* 100: 437-458.

Rizki, T.M. and R.M. Rizki. 1984. Selective destruction of a host blood cell type by a parasitoid wasp. *Proc. Natl. Acad. Sci. USA*, 81: 6154-6158.

Rizki, T.M. and R.M. Rizki. 1986. Surface changes on hemocytes during encapsualtion in *Drosophila melanogaster* Megen. In: *Haemocytic and Humoral Immunity in Arthropods*, pp. 157-190 (A.P. Gupta, ed.). John Wiley and Sons, New York.

Rizki, R.M. and T.M. Rizki. 1990a. Microtubule inhibitors block the morphological changes induced in *Drosophila* blood cells by a parasitoid wasp factor. *Experientia* 46: 313-315.

Rizki, R.M. and T.M. Rizki. 1990b. Parasitoid virus-like particles destroy *Drosophila* cellular immunity. *Proc. Nat. Acad. Sci.* 87: 8388-8392.

Rotheram, S.M. 1973a. The surface of the egg of a parasitic insect. I. The surface of the egg and first instar larva of *Nemeritis. Proc. Roy. Soc. Lond.* (B)183: 179-194.

Rotheram, S.M. 1973b. The surface of the egg of a parasitic insect. II. The ultrastructure of the particulate coat on the egg of *Nemeritis. Proc. Roy. Soc. Lond.* (B)183: 195-204.

Salt, G. 1963. The defence reactions of insects to metazoan parasites. *Parasitology* 53: 64-527.

Salt, G. 1965. Experimental studies in insect parasitism. XIII. The haemocytic reaction of a caterpillar to eggs of its habitual parasite. *Proc. Soc. Lond.* (B)162: 303-316.

Salt, G. 1968. The resistance of insect parasitoids to the defense reaction of their hosts. *Biol. Rev.* 43: 200-232.

Salt, G. 1971. Teratocytes as a means of resistance to cellular defense reactions. *Nature* 232: 639.

Salt, G. 1973. Experimental studies in insect parasitism XVI. The mechanism of resistance of *Nemeritis* to defense reactions. *Proc. Roy. Soc. Lond.* (B)183: 337-350.

Schmidt, O. and I. Schuchmann-Feddersen. 1989. The role of virus-like particles in parasitoid-host-interaction of insects. *Subcell. Biochem.* 15: 91-119.

Schmidt, O., K. Anderson, A. Will and I. Schuchmann-Feddersen. 1990. Virus-like particle proteins from the hymenopteran endoparasitoid are related to a component of the immune system in the lepidopteran host. *Arch. Insect. Biochem. Physiol.* 13: 107-115.

Schmidt, O. and N.A. Ratcliffe. 1977. The encapsulation of foreign tissue implants in *Galleria melonella* larvae. *J. Insect. Physiol.* 23: 175-184.

Söderhäll, K. and V.J. Smith, 1986. Pro-phenoloxidase-activating cascade as a recognitiion and defense system in arthropods. In: *Haemocytic and Humoral Immunity in Arthropods*, pp. 261-285 (A.P. Gupta, ed.). John Wiley and Sons, New York.

Stoltz, D.B. 1986a. Interactions between parasitoid-derived products and host insects: An overview. *J. Insect. Physiol.* 32: 347-350.

Stoltz, D.B. 1986b. Viruses in parasitic insects. In: *Fundamental and Applied Aspects of Invertebrate Pathology*, pp. 81-83 (R.A. Samson, J.M. Vlak and D. Peters, eds.). Fourth Internat. Colloq. Invert. Pathol. Wageningin, Netherlands.

Stoltz, D.B. and D. Guzo. 1986. Apparent haemocytic transformations associated with parasitoid-induced inhibition of immunity in *Malacosoma disstria* larvae. *J. Insect Physiol.* 32: 377-388.

Stoltz, D.B., D. Guzo, E.R. Belland, C.J. Lucarotti and E.A. MacKinnon. 1988. Venom promotes uncoating *in vitro* and persistence *in vivo* of DNA from a braconid polydnavirus. *J. Gen. Virol.* 69: 903-907.

Stoltz, D.B., P. Krell, M.D. Summers and S.B. Vinson. 1984. Polydnaviridae—a proposed family of insect viruses with segmented, double-stranded, circular DNA genomes. *Intervirology* 21: 1-4.

Stoltz, D.B. and S.B. Vinson. 1979a. Penetration into caterpillar cells of virus-like particles injected during oviposition by parasitoid wasps. *Can. J. Microbiol.* 25: 207-216.

Stoltz, D.B. and S.B. Vinson. 1979b. Viruses and parasitism in insect. *Adv. Virus Res.* 24: 125-171.

Streams, F.A. and L. Greenberg. 1969. Inhibition of the defense reaction of *Drosophila melanogaster* parasitized simultaneously by the wasp *Pseudocoila bochei* and *Pseudocoila mellipes*. *J. Invertebr. Pathol.* 13: 370-371.

Sweetman, H.L. 1958. *The Principles of Biological Control*. Wm. C. Brown Co., Dubuque, Iowa.

Takahashi, H., H. Komano and S. Natori. 1986. Expression of the lectin gene in *Sarcophaga peregrina* during normal development and under conditions where the defense mechanism is activated. *J. Insect. Physiol.* 32: 771-779.

Tanaka, T. 1987a. Effect of the venom of the endoparasitoid, *Apanteles kariyai* Watanbe, on the cellular defense reaction of the host, *Pseudaletia separata* Walker. *J. Insect. Physiol.* 33: 413-420.

Tanaka, T. 1987b. Morphological changes in the haemocytes of the host, *Pseudaletia separata*, parasitized by *Microplitis mediator* or *Apanteles kariyai*. *Dev. Comp. Immuncl.* 11: 57-63.

Tanaka, T. 1987c. Morphology and functions of calyx fluid in the reproductive tracts of endoparasitoid *Microplitis mediator*. *Entomophaga* 32: 9-17.

Tanaka, T. and H. Wago. 1990. Ultrastructural and functional maturation of *Apanteles kariyai*. *Arch. Insect Biochem. Physiol.* 13: 187-197.

Theilmann, D.A. and M.D. Summers. 1986. Molecular analysis of *Campoletis sonorensis* virus DNA in the lepidopteran host *Heliothis virescens*. *J. Gen. Virol.* 67: 1961-1969.

Vinson, S.B. 1972. Factors involved in successful attack on *Heliothis virescens* by the parasitoid *Cardiochiles nigriceps*. *J. Insect Physiol.* 18: 1315-1321.

Vinson, S.B. 1977. *Microplitis croceipes*: Inhibition of the *Heliothis zea* defense reaction to *Cardiochiles nigriceps*. *Exp. Parasitol.* 41: 112-117.

Vinson, S.B. 1990. How parasitoids deal with the immune system of their host: An overview. *Arch. Insect Biochem. Physiol.* 13: 3-28.

Vinson, S.B. and D.B. Stoltz. 1986. Cross-protection experiments with two parasitoid viruses. *Ann. Entomol. Soc. Am.* 79: 216-218.

Vinson, S.B. and D.L. Dahlman. 1989. Physiological relationship between braconid endoparasites and their hosts: The *Microplitis croceipes-Heliothis* spp. system. *Southewest Entomol.* (Suppl.) 12: 17-37.

Vinson, S.B. and J.R. Scott. 1974a. Parasitoid egg shell changes in a suitable and unsuitable host. *J. Ultrastructural Res.* 47: 1-15.

Vinson, S.B. and J.R. Scott. 1974b. Particles containing DNA associated with oocyte of an insect parasitoid. *J. Invertebr. Pathol.* 25: 375-378.

Walker, I. 1959. Die Abwehrreaktion des Writen *Drosophila melanogaster* gegen die zoophage Cynipidae *Pseudocoila bochei* Weld. *Rev. Suisse Zool.* 66: 569-632.

Webb. B.A. and M.D. Summers. 1990. Venom and viral expression products of the endoparasitic wasp *Campoletis sonorensis* share epitopes and related sequences. *Proc. Natl. Acad. Sci.* 87: 4961-4965.

Subject Index